Two Energy Futures:

A National Choice for the 80s

AUGUST, 1980

American
Petroleum
Institute

1980
American Petroleum Institute
2101 L Street, N.W.
Washington, D.C. 20037

Library of Congress Cataloging in Publication Data

American Petroleum Institute
Two energy futures.

Bibliography: p. 158
1. Power resources -- United States. 2. Energy Policy -- United States. I. Title.
TJ163.25.U6A45 1980 333.79'0973 80-24004
ISBN 0-89364-037-9
Printed in the United States of America

Table of Contents

Page

SUMMARY

The United States has the resources to face a bright energy future -- if appropriate decisions are made soon.

In 10 years, the nation could reduce oil imports by perhaps as much as 50 percent -- from the 1979 level of about 8 million barrels a day to around 4 million or 5 million barrels a day. Cutting imports by that amount would restore much of the energy security that the United States lost in the 1970s, when imports jumped dramatically. It also would help eliminate the conditions that created gasoline lines in some areas and led to sudden, steep oil price increases in 1973, 1974 and 1979.

The Challenge of the 80s: More Energy Security

If Americans start to take the right steps now, they can reduce dependence on foreign oil enough to have dramatic results by 1990 (other factors being equal).

The people then would have more certain supplies of energy for cars, homes, factories and jobs. Personal lifestyles would be more secure. The nation would have more stable energy prices than today, and its economy would be on a sounder footing. The dollar would be stronger than otherwise, the balance of payments improved, inflation lower and wages higher. Prices of imported consumer goods would be less than if present energy trends continue and fuel price increases would be restrained. America's international and foreign affairs could be more free from the threat of energy blackmail by foreign governments or terrorist groups. National security would be advanced. The nation could adjust more smoothly to a 1973-1974 oil embargo or 1979 Iranian oil export cutoff.

Additional domestic energy, produced and used efficiently, should also increase our economy's total output of goods and services, weaken OPEC's ability to raise world oil prices and make more foreign oil available to U.S. allies.

A Goal, Not a Prediction

The goal of reducing imports sharply by 1990 is not a promise, prediction or forecast of what <u>will</u> happen. Instead, it is a realistic description of what the nation's energy situation could be -- if the right actions are taken soon. The actions required are logical and feasible -- but not easy either for the political system or the energy producers. The nation, through its political process, may reject any or all of the energy opportunities it has. But the opportunities for significant change exist.

Cutting Oil Imports

An analysis of the U.S. energy situation shows that there are attainable, practical answers to our problems that require neither miracle solutions nor great self-sacrifice.

- The United States has vast energy resources -- far more than enough to support higher levels of domestic production.
- Given those resources, government policies will largely determine how much energy is actually produced over the next decade and beyond.
- With some adjustments in the rate of environmental progress and with concerted efforts to remove delays imposed by the government, considerably more domestic energy could be produced.
- Along with intelligent but not austere conservation, domestic energy production can be boosted to cover needed growth and also significantly reduce imports -- by as much as 50 percent.

A Possible World for 1990

By developing and using these domestic resources more effectively during the next decade, it is feasible to:

- stabilize oil and natural gas production at least close to today's levels;
- double our use of coal;
- at least triple the contribution of nuclear power; and
- get significant new production from synthetic fuels and renewable sources.

While the actual coal/oil/gas/nuclear/synthetic/renewable fuel "energy mix" could take a number of different forms by 1990, production ranges at such levels are reasonable -- if the nation chooses to make the required decisions.

But if the effort is not made, if we stay on today's course -- the course that is the basis of most current public forecasts -- instead of being down by as much as 50 percent in 1990, imports could increase by 25 percent or more . . . if that much foreign oil is available in world markets.

Policy Choices for the 1980s

The goal of reducing imports significantly does not assume any major technological breakthroughs. It is based on current technology, with continued technical progress, and it allows for near-term shortages of equipment or manpower. It does, however, assume definite changes in government policies:

- more energy from federal lands;
- a more careful balance between continued environmental progress and U.S. energy development;
- consistent government decisions that reduce uncertainty and encourage reasonable growth in nuclear power;
- increased reliance on the market system; and
- government promotion -- rather than preemption -- of private development of a commercial synthetic fuels industry.

Because of long lead times required to bring major new energy projects into production, the import goal for 1990 also assumes that such changes in government policy will be made soon.

Federal Lands

The rich energy resources on federal lands are indispensable in any effort to regain American energy security. The following comparisons suggest the importance of this potential.

Government lands, including the Outer Continental Shelf, now provide 16 percent of U.S. production of oil and natural gas liquids, 30 percent of the natural gas output and 8 percent of today's coal production. But studies show that they contain 37 percent of our undiscovered oil resources, 43 percent of undiscovered natural gas and 40 percent of our remaining coal.

Passage of many major laws controlling the use of land -- each intended for particular purposes -- has created a web of regulations that unnecessarily, and often unintentionally, work together to hold back energy development on these government lands. Administrative land withdrawals can also prevent energy development.

More exploration of federal lands will mean more energy. Safe and successful energy production in fragile areas such as the Gulf of Mexico and Alaska has shown that achieving energy security and preserving our natural heritage are compatible goals.

Environmental Regulation

The environmental laws adopted during the 1970s -- together with the efforts of concerned individuals and groups -- have moved the United States well along toward the goal of a cleaner environment. In the past 10 years:

- average sulfur dioxide levels in urban areas have dropped 30 percent;
- emissions of particulate matter (dust, soot, etc.) have declined by 40 percent nationwide;
- exhaust emissions of carbon monoxide and hydrocarbons from individual automobile models are down 80 percent to 90 percent; and
- hydrocarbon evaporation losses from gasoline tanks and carburetors of automobiles have been reduced by 85 percent.

We need not backslide on these and other environmental gains as we increase domestic energy supplies. Environmental and energy needs and goals can be balanced. We can continue to produce oil and natural gas, produce more coal, and begin to produce synthetic fuels -- and still enjoy a cleaner environment in 1990. Until now, rigidity rather than flexibility has often characterized the adoption and administration of environmental laws and regulations. As in the case of land use, this rigidity is reflected in more than a dozen major environmental laws enacted in the 1970s -- laws with commendable goals but restrictive provisions.

All of these have deterred or delayed the development of domestic energy resources.

The goals of these laws need not be changed. Without weakening the nation's environmental objectives, an environmental/energy balance can be achieved through more flexible government actions.

It is possible to increase coal use sharply in the next 10 years, for example, without damaging the environment. Although many political and economic problems remain to be solved, the basic technologies have been developed -- and are in use -- for mining, moving and using coal safely.

The causes and effects of other possible environmental problems -- any link between "acid rain" and coal use and the possible existence of a "greenhouse effect" -- are being carefully studied. The results of this research should, in the years ahead, help lead to sound decisions about even greater use of coal and other forms of energy.

Clean Air Act

The Clean Air Act standards and regulations especially pose problems for U.S. producers and users of energy resources: oil, natural gas, coal, synthetic fuels and shale oil.

Without retreating from the goal of improving the quality of the nation's air, Congress should modify the Clean Air Act to:

- ensure that the national air quality standards are based on sound, current scientific, medical and economic data;
- permit development of the promising energy potential that lies in areas which already meet national standards;
- allow companies to build new or more modern facilities in areas that have not yet attained the national standards;
- provide for realistic schedules and deadlines for achieving the national standards; and
- encourage the use of flexible, efficient concepts for improving air quality.

Nuclear Power

Most analysts believe that substantial gains can be made in nuclear power in the next decade if the government establishes consistent programs for measured nuclear growth. Among areas of attention are:

Enrichment services: The government is the only domestic operator of facilities that concentrate uranium for use in reactors. Plans to add to the enrichment capacity must be carried out if 1990 forecasts by nuclear experts are to be valid.

Storage of spent fuel: The administration has put forward a nuclear waste storage program that proposes research of potential repository sites, interim federal storage of excess waste, licensing of future waste facilities and creation of a State Planning Council.

Licensing: Congressional action is needed to consolidate siting procedures, to combine the processes for obtaining construction permits and operating licenses and to place limits on the reconsideration of issues previously decided.

Safety: While rigorous precautions have resulted in an impressive safety record, in the wake of Three Mile Island steps will be taken to upgrade crew training, improve operating procedures and audit day-to-day operations more closely.

Documentation of a National Energy Goal

Forecasts for 1990 by petroleum companies, government agencies, other institutions and individual academics indicate what they expect to happen. These forecasts differ, partly because of different assumptions of economic, environmental, technical and government factors. They reflect what each forecaster (or forecasting team) sees happening under those assumptions.

The trends suggested in these studies need not become the reality of the decade ahead. The nation can look instead at a 1990 situation built upon what could happen between now and then, given U.S. energy resources, if there are improved government policies. Pessimists -- and "realists" -- look at past trends and current policies and say, "No. The country will never make the choices needed to get there." The theory behind the 1990 accelerated case is that "people made the present policies; people can change them."

A comparison between the forecasts and the accelerated case numbers indicates that a national goal of reducing imports by as much as 50 percent is not an impossible romantic dream. The accelerated case is reasonable and reachable.

Following are the 1990 domestic crude oil (and natural gas liquids) production forecasts, published this year and last, which were used in this study:

FORECASTS OF 1990 PRODUCTION
CRUDE OIL[1]
(millions of barrels a day)

Department of Energy (mid case 4-79)	10.4
Professor Edward Erickson (best case)	10.0
Gulf Oil Corporation	10.0
National Petroleum Council	9.3-10.0
DOE Posture Statement (1-80)	9.0-10.0
National Academy of Sciences	7.5- 9.9
Data Resources Inc.	9.8
PIRINC	9.0
Shell Oil Company	8.6
Standard Oil Company of California	8.6
Arthur D. Little, Inc.	8.5
General Accounting Office	8.0
Exxon Company, U.S.A.	6.1

The natural gas forecasts reviewed for 1990 include individually published figures plus a group of projections from a survey of various organizations conducted and published in late 1979 by the American Gas Association:

FORECASTS OF 1990 PRODUCTION
NATURAL GAS[2]

(millions of barrels a day of oil equivalent)

	Lower 48 States	AGA Alaska Estimate	Total
*Gulf Oil Corporation	9.4	0.8	10.2
*General Accounting Office	8.3	0.8	9.1
*National Energy Plan II	7.3-8.3	0.8	8.1-9.1
*Mobil Oil Corporation	8.0	0.8	8.8
Nat. Academy of Sciences	-	-	4.9-8.5
DOE (mid case, 4-79)	-	-	8.2
*Atlantic Richfield Company	7.4	0.8	8.2
*Amoco Oil Company	7.4	0.8	8.2
Data Resources, Inc.	-	-	8.1
Nat. Petroleum Council	-	-	7.9
*Texas Eastern	7.1	0.8	7.9
Exxon Company U.S.A.	-	-	7.7
Texaco Inc.	-	-	7.6
Shell Oil Company	-	-	7.0
*Tennessee Gas Trans.	5.3	0.8	6.1

*Source: American Gas Association

Note the range of both sets of forecasts. For crude oil, about half are at approximately 10 mbd, and the rest are between 8 mbd and 9 mbd, with one significantly lower estimate. For natural gas, most are around 8 mbde, with a few above 9 mbde and only one as high as 10 mbde.

The 1990 Accelerated Case

The following table compares the current situation, the range of 1990 forecasts and the 1990 accelerated case proposed as a national goal. Again, it is not a promise, prediction or forecast of what will happen, but a realistic projection of what could happen if actions are taken soon.

UNITED STATES ENERGY[3]
(all numbers in mbd or mbde)

	1979	1990 Base Forecasts[a]	1990 Accelerated Case
Crude Oil/NGL	10.2	6.1 - 10.4	9 - 10
Natural Gas	9.1	6.1 - 10.2	8 - 9
Coal[b]	7.2	11.3 - 14.0	14.5 - 15.5[c]
Nuclear	1.3	3.1 - 4.2	4 - 4.5
Shale Oil	0	0.2 - 0.3	0.4 - 0.6
Renewables[d]	1.5	1.6 - 2.0	2 - 2.5
Consumption	36.8[e]	42.0 - 47.8	42 - 46
Oil Imports	8.2	7.5 - 11.3	4 - 5

[a]Range of recently published forecasts.
[b]Excluding exports of 0.8-1.0.
[c]Includes 0.1-0.3 liquid synthetics; 0.3-0.8 gas. Total synthetics from coal, oil shale, biomass are 1-1.5.
[d]Includes hydroelectric, solar, geothermal, biomass.
[e]Production plus imports do not equal consumption because of changes in stock levels and inclusion of product imports and NGL by volume, as commonly reported, rather than as crude oil equivalent.

A Goal That Is Reasonable, as Well as Attainable

So much emphasis is placed on oil imports that Americans tend to underestimate this nation's role as an energy producer. The United States now produces from its own resources 80 percent of the energy consumed by Americans. Cutting imports in half by 1990 would mean raising the share of energy produced at home by 10 percent in 10 years.

The increase from 80 percent to 90 percent is <u>reasonable</u>. Yet this relatively small shift toward greater self-sufficiency could provide major benefits to the American people. By acting wisely now and choosing soon to develop and use domestic resources more efficiently, Americans can, in 10 years, regain control of their energy future. By producing more U.S. energy to offset imports, and consuming all energy more wisely, Americans can put this nation's energy future back more securely into American hands.

Perspective

- Enormous efforts will be required by energy industries to reach even some of the lower numbers in the 1990 accelerated case. With oil and natural gas, the best apparent hope is that -- given the right decisions -- production can be stabilized <u>near</u> current levels.

- This paper speaks of a national goal of decreasing imports by perhaps as much as 50 percent by 1990 and cites an energy mix which would achieve that goal without environmental retreat or economic dislocation. But there is nothing magic about the percentage, the date or the mix. Clearly, <u>all</u> energy resources must be developed, and continued conservation is essential. The critical point is for the country to recognize the disadvantages which result from America's heavy dependence on imported oil, to understand that the current foreign domination is not inevitable and to move <u>now</u> in the direction of more U.S. control over its own energy affairs. If consumption turns out to be less than the range given for the goal, imports will be lower and the country will benefit from that decreased dependence. If the accelerated case proves to be overly optimistic even with improved government decisions or if consumption is higher than stated, it will be even more important in 1990 that a maximum effort had been made during the 1980s.

I. INTRODUCTION AND DOCUMENTATION

The 1950s, 1960s, and 1970s: Energy Security to Energy Dependence

America had secure supplies of energy during the 1950s and 1960s -- a time of unprecedented economic growth and rising standards of living in this country.

The 1970s, however, were a period of transition to grimmer times. Domestic oil and natural gas production began to fall. Our dependence on foreign oil rose rapidly -- from 3.4 million barrels a day in 1970 to 8.2 million barrels a day in 1979, or nearly half of U.S. petroleum consumption. The impact of a cutoff of part of those supplies was felt during the 1973-1974 embargo. As a result of limited supplies and federal price and allocation controls, gasoline became difficult to obtain.

Largely because of the new-found economic and political power of the oil-exporting countries, the cost of foreign oil was driven up sharply over the decade -- costs of driving cars, heating homes and running factories rose rapidly. The U.S. oil import bill increased from $3 billion in 1970 to $60 billion in 1979. And President Carter has predicted that it could reach $90 billion in 1980.[1]

Today, the availability of oil imports is subject to the goodwill and political stability of a few oil-exporting countries. It is also subject to the smooth passage of oil tankers across thousands of miles of exposed ocean and through many vulnerable "choke points" along their routes.

At any time, we could suddenly find ourselves paying dramatically higher prices for oil imports -- or find ourselves cut off from vitally needed supplies at any price. And -- even without a cutoff -- American consumers continue to send too many dollars abroad, and the United States suffers from reduced national security and a less flexible foreign policy.

The 1980s: A Decade of Choice

But our 1980 dependence on imports need not be a permanent feature of our lives. There are twin causes for our problems:

The United States uses too much energy and does not produce enough. The attainable, practical way out depends upon neither miracle solutions nor great self-sacrifice. It requires only that energy problems be approached with considerably more common sense so that more domestic energy can be produced and all energy can be used more efficiently.

We have great energy opportunities -- to seize or reject. We have a choice of two energy worlds for 1990:

- One is to continue down the energy road of the last several years. If we do, Americans will be at least as dependent on oil imports in 1990 as they are now. Given current expectations, foreign oil will account for about half of all the oil and more than 20 percent of all the energy we will be using at the end of the decade -- just as it does today.

- The other choice is for the nation to recognize that potentially there can be another and much improved 1990 energy world. Because of its physical and technological resources, the United States can reduce dramatically its dependence on foreign oil over the next 10 years, perhaps by as much as 50 percent. By 1990, foreign oil could account for perhaps 10 percent of all the energy we will be using -- just as it did during the 1950s and 1960s. Imports could be as low as 4 million or 5 million barrels a day -- if the right decisions are made and made soon.

Achieving this goal for the United States will not, of course, solve the energy difficulties that confront other oil-importing countries, including a number of our allies. Their problems do not have clear and available solutions. However, a more secure American situation would help them. The United States imports more oil than any other nation. To the extent that it curtails its demand for imports, it reduces pressure in the world oil market and the risks of import-dependency of other oil-importing countries.

Cutting oil imports roughly in half within a decade is not some impossible, romantic dream. Recent "best case" studies by energy specialists in government, business, industry and academia indicate that it is a reasonable and reachable goal for America.

This paper discusses the contributions which sensible people think can come by 1990 from various energy sources: oil, natural gas, coal, nuclear power, synthetic fuels, oil shale and renewable sources such as solar -- if the right decisions are made and made soon.

This introduction serves as a scene-setter, a documentation and a partial summary. Other chapters go into detail on national security, oil and natural gas, coal, nuclear energy, synthetic and renewable fuels, conservation, the environment, access to federal lands, the economic benefits of an improved 1990 energy position and what that improvement could mean to the individual. A list of sources consulted is also included.

Energy Resources of the United States

Our growing dependence on foreign oil during the 1970s led many people to conclude that the United States was running out of conventional sources of energy and that little or nothing could be done about it. But, although these sources are finite, they can last for decades more. In the case of coal, which can be converted to oil and gas, U.S. resources could last for centuries.

More specifically:

- Government and industry studies indicate that the United States can expect to produce about 150 billion barrels of crude oil and natural gas liquids in the future -- more than this country has ever produced. ("Natural gas liquids" are liquid petroleum that results from production and processing of natural gas.)
- The same studies indicate that remaining natural gas reserves plus future discoveries could total about 800 trillion cubic feet -- 40 percent more gas than the nation has ever produced and the equivalent of 142 billion barrels of oil.

These amounts of oil and gas are the equivalent of 40 years of production at current levels.

- Total recoverable reserves of coal are at least 250 billion tons (the equivalent of about 1,000 billion barrels of oil) -- more than three times greater than our recoverable resources of oil and gas.
- Recoverable reserves of uranium concentrate are about 700,000 tons -- enough to sustain the nuclear power industry well into the 21st century.

In addition, the United States has vast quantities of other fuels that could contribute increasingly to the nation's energy mix in the 1980s and beyond. For example, resources of relatively high-grade shale oil -- which will begin to make an appreciable contribution by 1990 -- are perhaps 80 billion barrels for initial commercial recovery and as much as 600 billion barrels for eventual recovery. This is twice as great as our oil and gas resources.

We can also ultimately expect abundant contributions from renewable sources, such as solar and geothermal energy and biomass. Commercial application of existing technologies and development of new technologies will likely make these sources increasingly competitive during the 1980s.

Beyond these energy sources, the United States has another tremendous resource -- conservation -- that is already reducing our dependence on imported oil. A few years ago most people thought of conservation as simply "doing without," and many experts believed that there was little opportunity to reduce energy consumption without economic hardship. The experience of the past few years -- as consumers have adjusted to rising prices -- has dramatically altered this belief.

Today, Americans are using energy far more efficiently than in the past. Economic growth now requires less additional energy consumption than it required 10 to 15 years ago. Existing and new technologies that further increase energy efficiency will become more competitive in the years ahead.

In brief, the United States has ample resources to meet its energy needs. The issue is how effectively the nation will choose to use them.

Outlook for 1990: The Present Course or the Attainable Goal?

Good use must be made of these vast resources in the years ahead in order to sustain a reasonable rate of economic growth -- perhaps 3 percent annually. Even better use must be made of them if we are also to attain the goal of significantly reduced oil imports by 1990. As noted, conservation will make a major contribution. However, even with the improved levels of conservation that we have seen in the last year or so, considerable new supplies of energy will be required to satisfy the nation's future energy needs. Recent forecasts by some who advocate and expect substantial additional conservation still indicate that the United States will need 15 percent to 20 percent more energy in 1990 than it is using today. In short, clearly there is no acceptable way of "saving our way out."

But by acting wisely soon and choosing to develop and use domestic resources more efficiently, we could:

- halt the decline in U.S. oil production at or near current levels and hold production at about 9 million to 10 million barrels a day;

- stabilize natural gas production fairly close to today's level by holding it at the equivalent of about 8 million to 9 million barrels of oil a day;

- double our use of coal by 1990 to about 14.5 million to 15.5 million barrels a day of oil equivalent;

- at least triple the contribution of nuclear power by 1990 to 4 million to 4.5 million barrels a day of oil equivalent; and

- get significant new production from synthetic fuels and renewable sources -- increasing total production to perhaps 3 million to 4 million barrels a day of oil equivalent.

Production of domestic energy at those levels -- along with sensible but not austere conservation -- could cut our oil imports by roughly one-half by 1990 to about 4 million or 5 million barrels a day.

Of course, the actual "energy mix" doubtless would turn out to be somewhat different than given above. We would end up with more than suggested from some energy sources and less from others. No one can look down a 10-year road with precision. The important point is that the production ranges cited are reasonable, if the nation chooses to make the required decisions.

However, if the public and its government do not take the necessary steps, if policy changes are not instituted, if we consider OPEC to be unalterably in control of our lives, if we erroneously believe there is no way out, this reduction in imports will not be achieved. Given current expectations -- based on past trends and only slight improvement in present government policies -- in 1990 there is likely to be:

- less oil production;

- less natural gas production;

- slower growth in coal production;

- far more uncertain growth in nuclear power; and

- much less new production of synthetic fuels.

If the nation follows that kind of course, if it allows current expectations to become reality, by 1990 our imports could increase by 25 percent or more -- if that much oil were to be available in world markets.

The results of cutting imports by as much as 50 percent would be dramatic. Americans would once again have secure supplies of energy for cars, homes, factories and jobs. We would have more stable energy prices than we have today. The national economy would be on a sounder footing. Our internal and foreign affairs would be more free from the threat of energy blackmail by foreign governments or terrorist groups. We could adjust more readily to a 1973-1974 oil embargo or 1979 Iranian oil export cutoff.

To accomplish so much in the face of some recent gloomy energy forecasts for 1990 might suggest that the nation would have to risk a single-minded national effort that would force us to abandon our other national goals. But pursuing this course would not, in fact, require drastic new measures. Shifts in policy would be necessary, but not changes in our values -- given our resources.

It would simply require that, by 1990, the United States produce 90 percent of the energy that it needs, rather than the 80 percent that it produces now -- the 80 percent it would produce in 1990 if we stay on our present course. Although extremely difficult, an increase from 80 percent to 90 percent is achievable.

Policy Choices for the 1980s

There would be problems at each step in reaching this goal, which would be neither cheap nor easy to attain. But there also would be sensible solutions to those problems.

The import reduction goal does not assume any major technological breakthroughs -- only continued technical progress. It takes into consideration potential physical problems, such as near-term shortages of equipment and manpower. It assumes basically that unnecessary government regulatory roadblocks, red tape and delays will be substantially reduced and that the market will have more opportunity to choose the most economic options. And it assumes that, because of the long lead times required to bring major new energy projects to production, these improvements in government policy will be made soon.

More specifically, the goal assumes that the following actions will be taken:

- The federal government will conduct an aggressive leasing program that permits energy companies access to coal deposits and lands where oil and gas may be found. There must be government permission for more energy exploration of federal lands, including the Outer Continental Shelf (OCS).

- The government will inject a more reasonable balance into environmental rules, relaxing overly strict standards where necessary to develop energy but always protecting public health with an adequate margin of safety.

- The government will adopt consistent policies that reduce uncertainty and encourage reasonable growth in nuclear power.

- We will increase our reliance on the market system, including decontrol of energy prices, as planned. (The so-called "windfall profits" tax, however, is assumed to remain in effect.)

In short, the goal -- which, again, is not a prediction -- assumes that we will balance our national needs, make our rules reasonable and remove unnecessary roadblocks.

Energy from Federal Lands

The federal government owns about one-third of all the land in the United States and all of the OCS beyond state jurisdiction. The rich energy resources of these areas are indispensable in any effort to regain American energy security. The following comparisons suggest the relative size of the potential involved.

Federal lands, including the OCS, now provide 16 percent of U.S. production of oil and natural gas liquids and 30 percent of the nation's natural gas output. Yet, government studies indicate that these lands contain 37 percent of our undiscovered oil and natural gas liquids and 43 percent of our undiscovered natural gas.

Federal lands also hold 40 percent of the nation's remaining coal supplies -- but provide only 8 percent of production.

And, finally, federal lands hold 80 percent of our recoverable western oil shale.

In short, these federal areas hold many of the resources that the nation will need if it is to meet the 1990 goal.

However, laws and regulations unnecessarily -- and often unintentionally -- foreclose or delay exploration of most of these areas and deprive Americans of vast energy resources. In addition, administrative land withdrawals can have the same effect. Careful exploration of these areas can and should be expanded and accelerated so that these resources can contribute their full potential to the goal of energy security.

We can do this and protect our natural heritage. Past development of energy resources on federal lands has demonstrated that these two essential American goals -- energy security and protection of that natural heritage -- can be compatible. We can and should achieve an effective balance between them.

Environmental Regulation

Another important national goal during the past decade has been a cleaner environment. This goal has been pursued vigorously, and much progress has been made. However, that vigorous pursuit has sometimes limited development of domestic oil, natural gas and coal. And, unquestionably, it has sometimes slowed progress toward restoring our energy security.

Often unintentionally, environmental laws and regulations have created a procedural maze that can delay virtually any major energy project indefinitely. Those regulations can and should be streamlined so that domestic energy sources can be developed more rapidly. In addition, some environmental standards are unrealistically strict. These standards can be relaxed somewhat to permit more effective use of our resources -- while still reducing health risks for the American people.

Clean Air Act standards and regulations especially pose problems for U.S. producers and users of oil, natural gas, coal, synthetic fuels and shale oil.

Congress should clarify aspects of the Clean Air Act to ensure:

- that the national air quality standards (some of which are based on studies done more than a decade ago) reflect the wealth of scientific and medical information on air pollutants collected over the past 10 years.

- that energy development is permitted in those areas of the country that presently meet the national air quality standards. These areas have the greatest potential for energy development, but development of them is hampered by a complex and restrictive set of regulations.

• that companies are allowed to build new facilities in areas of the country that have not yet achieved the air quality standards. Development in these areas is hampered by a separate set of regulations. The current restrictive policy for these areas (called the "offset" policy) should be replaced with a policy that allows expansion and modernization of energy facilities, while requiring tight emission controls.

• that realistic schedules and deadlines are established for achieving the standards, replacing the present rigid deadlines. Presently, if these rigid deadlines are not met, the federal government can cut off funds and freeze energy development and industrial expansion in states that fail to meet the standards.

Such changes would <u>not</u> be a retreat toward dirty air. Rather, they would provide flexibility in the speed with which we move toward our environmental goals. The urgency of national needs changes from year to year and from decade to decade, and so do our national priorities. We must be flexible enough to respond to those changes. We can develop more domestic energy <u>and</u> continue to improve the quality of our environment by balancing our efforts to achieve both goals.

<u>Nuclear Power</u>

The third key area of government policy concerns nuclear power. The United States developed the basic technology of the nuclear power industry and committed itself to measured growth of that industry. However, recent controversy has cast doubt on that commitment. Our pace has slowed dramatically, while other nations -- such as Great Britain, France, West Germany, Sweden, Japan and the Soviet Union -- have moved ahead at a steady pace.

Safety, procedural and waste disposal problems facing American nuclear power can be overcome. Safety precautions can and will be strengthened. Many unnecessary procedural hurdles, which create lead times twice as long as those in other countries, can be removed. These changes can speed construction of a new nuclear power plant by several years.

Nuclear experts have determined that nuclear fuel waste should be stored in underground geologic formations. The president has announced a comprehensive radioactive waste management program. The program calls for an extensive research and development effort to select a repository site; a government contingency program for the storage of spent fuel and disposal of low-level nuclear waste; licensing by the Nuclear Regulatory Commission of any future nuclear waste management facilities; and creation of a State Planning Council to advise the executive branch on nuclear waste policy.

The nation can renew its commitment to nuclear power -- and can do so in a safe and rational manner. Reaching the national goal for 1990 will require completion of nuclear plants that now have government construction permits, and virtually all studies point out that continued improvement of our energy situation after 1990 will require further growth of nuclear power.

Other Government Policies

Aside from positive action in these three specific areas, there must be more consistency in government policies across the board. It will not be possible to gain greater energy security unless government policies are consistent, both with each other and from year to year.

In addition, government can play other specific roles. For example, its help can accelerate development of a commercial synthetic fuels industry. In such efforts, the nation can move faster toward the 1990 goal if government promotes rather than pre-empts private development. At this critical time, the nation will progress more surely if it takes full advantage of the inherent efficiency of the private sector and of the flexibility and responsiveness that thousands of competing private companies can offer.

Documentation of a National Energy Goal

Two basic sets of numbers are used throughout this discussion of 1990. To prevent misunderstanding, it is essential that the two are not confused. (The sources consulted in preparing the sets of numbers are identified on pages 158-160.)

One set consists of published 1990 forecasts by petroleum companies, government agencies, individual academics and other institutions. These reflect what the forecasters think will happen. Some forecasters give a range -- low, middle and high cases. Others offer just one number. Every forecast is different from every other one -- because each was made at a different time and because each used different assumptions about economic, environmental, technical and governmental factors.

For example: Gulf Oil Corporation forecasts 10 million barrels a day of domestic oil production in 1990;[2] Exxon

Company, U.S.A., forecasts 6.1 million barrels a day.[3] And many other forecasts fall in between. This paper merely reports the forecasts; it makes no effort to reconcile them.

Of course, none of these forecasters knows precisely what U.S. energy production levels will be in 1990. With oil and gas, for example, no one can know without additional exploration and drilling.

Another set of figures is used throughout this discussion to describe the 1990 "accelerated case" that could exist in a different energy world. These numbers are not forecasts or predictions, and do not represent what will happen in 1990, or might happen, or is likely to happen. Rather, they reflect what appears to be achievable, given the resources of this nation, if the right decisions are made and made soon.

No rational person believes that federal and state governments will suddenly reverse policies of the last decade and move with lightning speed to a "perfect" energy course.

But clearly there will be much more movement if the people are determined to improve their energy condition than there will be if the people think they must accept the inevitable. There can be more movement than planners think there will be. Thus, the accelerated case described for 1990 -- the case that could cut imports up to 50 percent -- is drawn from the possible, not from the anticipated.

That accelerated case, a composite of "high supply" projections, assumes much more efficient use of our energy resources. It suggests that the public can become involved with the country's energy situation, that informed Americans can then opt for government policies designed to remove obstacles to domestic energy production and that such people can soon influence government to make those enlightened decisions.

Yet the accelerated case is not "blue sky." It is reasonable and reachable over a 10-year period. It contemplates no technological breakthroughs, no environmental retreat and no legislative action which could be considered unrealistic from an energy-conscious Congress.

Because the forecasts and studies produce ranges, the accelerated case in this paper for each energy source is also presented as a range. Its upper end represents the estimates of analysts who have concentrated on an accelerated case, as well as some of the higher forecasts of what is expected to happen. The lower end of the range recognizes that there are some low forecasts.

The great range shown in the forecasts is evidence that the nation cannot plan rigidly on the basis of any particular energy projection -- whether pessimistic or optimistic. The wise course is to take necessary steps to determine what our domestic energy resources are and to maximize the economical production of them.

Great efforts will be required to achieve small improvements in production. If the production of some sources fails to reach the levels of the accelerated case, the nation will still be better off in 1990 than it otherwise would have been. And the 1990 import goal might still be met by higher production of other energy sources or increased conservation.

Following are the details supporting the accelerated production levels for oil, natural gas, coal, nuclear power, synthetic fuels and renewable energy sources. Also included are the details on expected energy consumption.

While this treatment concentrates on 1990, it must be recognized that there is nothing unique about that date. Progress toward the goal and beyond should be continual.

Oil and Natural Gas

For well beyond the coming decade, oil and natural gas will be principal sources of energy in the United States. The issue is how large a contribution they will make to a national goal of reducing imports by as much as 50 percent by 1990. The size of that contribution will depend in large part on what government decisions are made.

For many years, the nation has been using considerably more domestic oil and natural gas than it has been finding. Production has been declining.

Domestic oil and natural gas liquids production peaked at 11.3 million barrels a day (mbd) in 1970 and declined to about 10 mbd in 1979. Natural gas production peaked at about 11 million barrels a day of oil equivalent (mbde) in 1973 and declined to about 9 mbde in 1979.

The domestic oil decline has been slowed by the advent of Alaskan North Slope oil, which began flowing through the trans-Alaska oil pipeline in 1977. North Slope production rose to 1.5 mbd in late 1979. Further discoveries will be required not only to increase North Slope production, but also to sustain production rates over the long term.

While recent exploratory results in such areas as the Gulf of Alaska and the Baltimore Canyon off the New Jersey coast have not been encouraging, it is believed that most large new domestic discoveries of oil and gas are still likely to be in the western states, in Alaska and on the Outer Continental Shelf (OCS).

Under present conditions, lead times between initial exploration and first production in some areas are estimated to be 8 to 11 years. Thus, rapid exploration and development of the oil and gas resource potential of the OCS and other areas are essential to help offset declining onshore production. The rate of production growth in these areas will depend largely on governmental leasing and regulatory policies.

As indicated, the accelerated case in this paper assumes that those government policies will be improved.

The forecasts reviewed for this paper show a 1990 range of 6.1 mbd to 10.4 mbd of crude oil and natural gas liquids production.

For natural gas, they show a 1990 range of 12.5 trillion to 20.9 trillion cubic feet a year (6.1 mbde to 10.2 mbde).

Following are the 1990 domestic crude oil production forecasts, published this year and last, which were used in this study. They are from oil companies, academics, the government and such independent sources as Arthur D. Little, Inc. and the National Academy of Sciences.

FORECASTS OF 1990 PRODUCTION
CRUDE OIL & NATURAL GAS LIQUIDS[4]

(millions of barrels a day)

Department of Energy (mid case, 4-79)	10.4
Professor Edward Erickson (best case)	10.0
Gulf Oil Corporation	10.0
National Petroleum Council	9.3-10.0
DOE Posture Statement (1-80)	9.0-10.0
National Academy of Sciences	7.5- 9.9
Data Resources Inc.	9.8
PIRINC	9.0
Shell Oil Company	8.6
Standard Oil Company of California	8.6
Arthur D. Little, Inc.	8.5
General Accounting Office	8.0
Exxon Company U.S.A.	6.1

The natural gas forecasts reviewed for 1990 include several from a survey of various organizations taken and published in late 1979 by the American Gas Association. The others are also public figures.

FORECASTS OF 1990 PRODUCTION
NATURAL GAS[5]

(millions of barrels a day of oil equivalent)

	Lower 48 States	AGA Alaska Estimate	Total
*Gulf Oil Corporation	9.4	0.8	10.2
*General Accounting Office	8.3	0.8	9.1
*National Energy Plan II	7.3-8.3	0.8	8.1-9.1
*Mobil Oil Corporation	8.0	0.8	8.8
Nat. Academy of Sciences	-	-	4.9-8.5
DOE (mid case, 4-79)	-	-	8.2
*Atlantic Richfield Company	7.4	0.8	8.2
*Amoco Oil Company	7.4	0.8	8.2
Data Resources, Inc.	-	-	8.1
Nat. Petroleum Council	-	-	7.9
*Texas Eastern	7.1	0.8	7.9
Exxon Company U.S.A.	-	-	7.7
Texaco Inc.	-	-	7.6
Shell Oil Company	-	-	7.0
*Tennessee Gas Trans.	5.3	0.8	6.1

*Source: American Gas Association

Note the ranges of both sets of forecasts. For crude oil, about half are at approximately 10 mbd, and the rest are between 8 mbd and 9 mbd, with one lower estimate. For natural gas, most are around 8 mbde, with a few above 9 mbde and only one as high as 10 mbde.

One of the few published studies that looks specifically at a "best case" for oil production was done by Edward Erickson, professor of economics and business at North Carolina State University. Erickson made "an estimate of what the supply response for U.S. crude oil production might be in an environment which is relatively unconstrained in certain important ways," including price and access to government lands. Erickson's case, which does <u>not</u> include a "windfall profits" tax, yields a crude oil production rate of 10 mbd from 1985 through

1990.[6] The author does not make an estimate for natural gas liquids. If natural gas liquids are added and the effects of the tax are subtracted, Erickson's case would remain about the same: 10 mbd.

Another study that looks at relatively unconstrained cases is the report released in December 1979 by the National Academy of Sciences. This report examines "enhanced supply" and somewhat more aggressive "national commitment" scenarios for all energy sources. The assumptions are quite similar to those of the accelerated case in this paper. For oil and natural gas liquids, the National Academy of Sciences' best case projects 9.9 mbd in 1985 and 1990.[7] Much of the work for this study was done some time ago, so the specific assumptions on prices and taxes do not include recent developments. Higher prices would raise the projection, and higher taxes would lower it. The net effect of these developments on the Academy's results is unclear.

Gulf Oil Corporation's most probable case for 1990 crude oil and natural gas liquids production is 10 mbd, including the "windfall profits" tax and current -- not accelerated -- federal leasing policies.[8]

In December 1979 the National Petroleum Council (NPC) published a survey of supply and demand data from 20 organizations. The average 1990 case in this survey was 10 mbd of crude oil and natural gas liquids and 16.2 trillion cubic feet per year, or 7.9 mbde, of natural gas. Tentative results from an updated NPC survey show an average case of 9.9 mbd for crude oil and natural gas liquids, with a low case of 9.3 mbd.[9]

The Department of Energy (DOE) 1979 projections for domestic crude and NGL production show a high of more than 13 mbd in 1990, although this number depends on the high end of the United States Geological Survey reserves estimate. A middle case, based on a more likely reserves estimate, is about 10.4 mbd.[10] A January 1980 DOE "posture statement" notes that oil production "is expected to remain near current levels of 9 to 10 million barrels per day through the year 2000."[11]

Some forecasts of what is expected to happen are substantially lower. Exxon Company, U.S.A., with the lowest oil company forecast, sees 6.1 mbd in 1990.[12] Shell Oil Company predicts 8.6 mbd in 1990.[13] However, Shell has recently testified that "the opportunity exists to find and produce significantly greater volumes of oil from the Alaskan OCS through a concerted, cooperative effort of government and industry."[14] Shell estimated that an accelerated leasing schedule could result in 4 mbd of Alaskan OCS production in 1995, versus only 1 mbd by that date under the currently proposed Interior Department schedule.[15]

In sum, the accelerated case suggests the possible 1990 production of some 9 mbd to 10 mbd of crude oil and natural gas liquids and some 16.5 trillion to 19 trillion cubic feet of natural gas a year (8 mbde to 9 mbde). Production at these rates could stablize both oil production and natural gas production near current levels.

Coal

There is widespread agreement that coal use can and must increase over the coming decade. The size of the increase is a much more difficult question, although a supportable best case shows that coal use can double.

In 1979, the United States consumed about 680 million tons of coal, or 7.2 mbde. A National Coal Association forecast for 1990 shows a mid to upper range of about 1.2 billion to 1.45 billion tons, or about 12.5 mbde to 15.5 mbde.[16] DOE's best 1990 case for coal use is about 14.5 mbde.[17] The National Academy of Sciences' report shows best case production of 15.3 mbde;[18] assuming about 1 mbde of coal exports, U.S. consumption would be about 14.3 mbde.

Therefore, a doubling of coal use by 1990 appears to be reasonable for the accelerated case, which assumes energy-conscious government decisions.

Nuclear Energy

Defining an accelerated case for nuclear power generation is particularly difficult, given the intense emotion that has surrounded the issue over the past few years and the political impact of the accident at Three Mile Island in 1979. With timely siting, permitting and construction, huge increases in nuclear power are possible -- and huge increases were predicted in the early 1970s. Even after problems and longer-than-anticipated lead times began reducing that optimism, forecasts in 1975 and 1976 showed nuclear power quadrupling by 1985 and increasing seven-fold by 1990.

Forecasts made later, in 1977-78, still projected strong growth but substantially less than earlier -- closer to a four-fold increase by 1990. As of mid-1980, few forecasts had been published which take into account the findings of the Presidential Commission on the Accident at Three Mile Island and the effects of that accident. As would be expected, however, discussions with nuclear analysts in DOE and the nuclear and electric power industries indicate even more pessimism.

The current forecasts generally limit nuclear growth to those plants already under construction or in permit review. This growth would roughly triple nuclear energy by 1990, from 1.3 mbde today to around 4 mbde in 1990. Little support now can be found for a 1990 case reflecting the earlier optimism. The huge 1990 growth rates projected earlier cannot now be realistically considered. Furthermore, lead times are such that even a rapid change in policies may add relatively little growth by 1990. (Such a change could, however, have a large effect beyond 1990. The National Academy of Sciences, in fact, states that "effective new government policies will be required if nuclear energy is to continue to grow past about 1990.")[19]

Considering only the more recent data -- not the forecasts of the early and mid-1970s -- a tripling of nuclear capacity and power generation by 1990 seems to be a realistic expectation. A DOE forecast that was updated in October 1979 shows an increase in installed capacity by a factor of about 2.5 to 3.[20]

The Atomic Industrial Forum's latest projection shows an increase in installed capacity by a factor of 3.2. The factor would be around 3.4 if all 110 reactors under construction and ordered could be completed by 1990.[21] (This group of reactors includes 89 with government construction permits, two more with "limited work authorizations" and 19 on order.)

Figures from the current estimate of the National Electric Reliability Council (NERC), a utility group, show an increase in capacity by a factor of about 2.9 by 1989 and an increase in actual generation by a factor of 3.4 for the same period.[22]

Overall, 72 nuclear plants are now licensed for operation, and by 1990 the forecasts see the completion of another 100 or so. As indicated, all of these are now under construction or ordered.

The National Academy of Sciences' study considered a case with "strong government commitments" and with the nuclear industry building plants at full capacity. This case, recommended by the Academy, shows capacity increasing by a factor of 4.2, which implies about 35 new nuclear plants beyond those already identified at the time the report was published. An even stronger case, with commitment to breeder reactors, gave a capacity-increase factor of 4.6.[23] However, there is little additional support for cases as optimistic as these, and these numbers are not reflected in the 1990 accelerated case of this paper.

In sum, analysts appear to agree that nuclear capacity will about triple by 1990. Assuming that operating factors remain approximately the same, actual power generation would also triple to about 4 mbde. At best, if all plants now under permit or ordered were allowed to become operational by 1990, the figure could increase to about 4.5 mbde.

This higher figure, though, is uncertain. The situation is still unsettled following Three Mile Island, and the longer term effects on governmental policy are not yet clear. While the government may facilitate nuclear plant construction, unquestionably there is still a chance that some plans for new plants will slip past 1990.

With that qualification, a reasonable accelerated case is considered to be about 4 mbde to 4.5 mbde. This range actually represents little step-up over current forecasts. The important improvement in the accelerated case is that the large government policy uncertainties would be resolved, so that the nation would be much more confident of reaching these production levels. Without clarification of government policy, the uncertainties could cause nuclear energy to fall far short of the projections.

The total accelerated energy case for 1990 could tolerate some slippage in new nuclear power. Given a reasonable amount of lead time, there should be "room" to shift somewhat toward a higher proportion of new coal-fired plants, if necessary. Of course, if nuclear power experiences major unforeseen problems -- such as a renewed, lengthy moratorium -- the 1990 import goal would not be met.

Synthetic Fuels

With synthetic fuels, the accelerated case used for this paper is not necessarily on the high side of today's estimates. Virtually everyone is optimistic about synthetic fuels for the more distant future. This is one of the most challenging areas of development over the next 30 years. However, there are many unknowns today, and an overly ambitious program launched before there is greater knowledge might commit the nation to alternatives that are not cost-effective and that rule out better choices. Thus, extremely optimistic estimates for various synthetic fuels were consciously avoided in laying out reasonably achievable production levels for 1990.

The Energy Security Act, passed in June 1980, sets synthetic fuels production goals of 0.5 mbde by 1987 and 2 mbde by 1992. The policy scenarios recommended by the National Academy of Sciences project slightly more than 2 mbde in 1990.[24]

Various published government and industry data show potential production of individual synthetic fuels as follows: liquids from coal, 0.1 mbd to 0.3 mbd; gas from coal, 0.3 mbde to 0.8 mbde; shale oil, 0.4 mbd to 0.6 mbd; biomass conversion, 0.2 mbde to 0.3 mbde.

Thus, for the national 1990 import goal of this paper, a total of 1 mbde to 1.5 mbde was taken as a reasonable and supportable case for synthetic fuels production. Such a range allows for adjustments as more is learned about synfuels production.

Solar, Hydroelectric, Geothermal Energy

The remaining energy sources -- solar, hydroelectric and geothermal -- are relatively small, both in current and forecast 1990 production. Most forecasts group them in a single category.

These sources currently supply about 1.5 mbde, mostly from hydroelectric power. Several private and government forecasts indicate an increase of around 0.3 mbde to 0.8 mbde by 1990, for a total of 1.8 mbde to 2.3 mbde. The National Academy of Sciences' estimates are somewhat higher, with a recommended case that would yield somewhat less than 3 mbde and a highest case of around 4 mbde.[25]

A consensus, including the National Academy of Sciences, suggests an increase of about 0.5 mbde to 1 mbde by 1990, for a total of 2 mbde to 2.5 mbde. This paper uses that range for the 1990 accelerated case. Achieving the full potential of solar energy will require technological improvements to reduce current costs, and forecasters expect that a significant solar contribution will come later than 1990.

Consumption

Forecasts of U.S. energy consumption levels have been dropping steadily over the past several years. Conservation from higher prices and other factors has been a significant part of virtually all recent studies and very strong in several. Thus, the 1990 best case in this paper for conservation is close to the current forecasts for that date. These levels can be achieved without dramatic changes in American lifestyle. Significant energy conservation beyond current forecasts would appear to depend on sharply higher prices or government mandates or subsidies that go far beyond those already in effect.

The resulting U.S. energy consumption levels for the 1990 accelerated case range from 42 mbd to 46 mbd of oil equivalent, compared with 1979 consumption of about 37 mbde.

It is important to re-emphasize what is happening to energy use in the United States. Exxon, among the most conservative of the oil companies on production forecasts, sees one of the smallest increases in consumption, projecting only 43 mbde in 1990. The company notes that this is 16 mbde less than "would have been expected if energy prices and consumption patterns were assumed to follow the trends of the 13-year period (1960-1972) immediately prior to the 1973 Arab embargo."[26]

Shell's more recent forecast is even lower, at 42 mbde.[27] These numbers compare with estimates of 50 mbde to 55 mbde for 1990 made only two to five years earlier.

DOE's projections made in the spring of 1979 show a lowest case consumption of about 45.5 mbde in 1990,[28] although the 1980 DOE posture statement notes that "energy consumption will be held near current levels over the next five years" and after that will grow more slowly than in the past.[29]

The recent Harvard Business School report, Energy Future, projects an increase in consumption to about 46 mbde in 1990, although the report's authors have since written that "at the very least, our aim should be zero energy growth for the 1980s." They state that this aim can be met through "productive conservation" with "positive economic growth." That is, with the right investments, we would not face factory shutdowns, higher inflation and so on. They cite as requirements the "removal of institutional barriers" and federal subsidies covering 40 percent to 60 percent of investments.[30]

One final point on consumption is worth noting. Accelerated use of coal could tend to make total energy consumption somewhat higher than it would be if coal were to provide a smaller proportion of our total energy. The reason is that coal is more "energy intensive" than oil or gas. That is, more energy is generally needed to produce coal, transport it and operate the necessary pollution control equipment when it is used. Also, producing synthetic fuels from coal necessarily involves some conversion losses. So, slightly higher consumption would not mean less conservation; it would simply mean a slightly different energy mix.

Import Reduction Steps

In these accelerated cases for domestic energy production and consumption, the import balance sheet would be affected in the following ways:

- With production of oil and natural gas at best stabilized near 1979 levels, increasing imports would no longer be necessary to make up for declining production. This step depends on the fact that current forecasts do not expect total U.S. oil consumption -- production plus imports -- to increase during the 1980s.

- The increased coal production would permit coal to displace up to 2.5 mbde of oil and natural gas in utilities and industry (as explained in Chapter IV), with the gas then used elsewhere in place of oil.

- New coal-fired power plants, nuclear power and renewable sources (hydroelectric, solar and geothermal) will meet increasing demand for electricity, so that this higher demand will not require additional oil imports.

- Production of 1 mbde to 1.5 mbde of synthetic fuels from coal, oil shale and biomass will reduce imports.

Under that schedule, 1990 oil imports could be reduced by 3 mbd to 4 mbd beneath what they would otherwise be. As noted, this reduction could cut imports by perhaps as much as half from their 1979 level of slightly more than 8 mbd to a figure closer to 4 mbd -- assuming that the trend toward energy conservation continues as forecast.

Reduction of imports will be aided if refiners are able to install new technologies that convert a higher proportion of domestic residual oil into gasoline and other light products. These additional light products would help replace imported oil.

This possibility has not been counted upon on a large scale for the 1990 accelerated case because significant help from it would involve major new design and construction. A nationwide program to install such new technologies might encounter equipment and manpower limitations if attempted in addition to an ambitious synthetic fuels program.

To a lesser extent, refiners may be able to use existing equipment to convert more heavy oil. Flexibility in operation may allow a shift toward slightly more production of light products. Although the heavy oil is currently needed in utility and industrial boilers, this step could be taken when more coal becomes available to replace that oil.

The table on page 23 summarizes domestic energy production and consumption. It compares the current situation, the expected 1990 situation and the accelerated case which is reasonable for 1990 if government policies are soon improved and steps are taken to implement them.

DOE Study

Finally, further support for the 1990 accelerated case comes from a January 1980 report sponsored by the Department of Energy. The report, done by the Engineering Societies Commission on Energy Inc. (ESCOE), looked at a similar goal but took the reverse approach. It began by defining the goal and then investigated how the nation could achieve it.

The ESCOE goal was defined by federal law and policy -- the Fuel Use Act, National Energy Plan II and the subsequent new energy initiatives described by the president on July 15, 1979. The objective, stated by the president, is to reduce oil imports to 4.5 mbd -- about 45 percent below 1979 levels -- by 1990.[31]

ESCOE concluded that "the basic objective of significantly reducing oil imports by 1990 is achievable within reasonable national economic assumptions if the necessary decisions are made and actions taken now."[32] ESCOE's scenario is nearly identical to the accelerated case described in this paper.

Specifically, ESCOE's scenario included:

- increasing coal use 7.6 percent per year to about 16 mbde in 1990;
- stopping the decline in oil and gas production;
- bringing all nuclear plants with construction permits into operation;
- producing 1.5 mbde of coal synfuels; and
- growth of total U.S. energy consumption to about 46.5 mbde in 1990.[33]

The sole purpose of ESCOE's study was to see whether the administration's target of reducing imports to 4.5 mbd is

UNITED STATES ENERGY[34]

(all numbers in millions of barrels per day of oil equivalent)

	1979	1990 Base Forecasts[a]	1990 Accelerated Case
Crude Oil/NGL	10.2	6.1 - 10.4	9 - 10
Natural Gas	9.1	6.1 - 10.2	8 - 9
Coal[b]	7.2	11.3 - 14.0	14.5 - 15.5[c]
Nuclear	1.3	3.1 - 4.2	4 - 4.5
Shale Oil	0	0.2 - 0.3	0.4 - 0.6
Renewables[d]	1.5	1.6 - 2.0	2 - 2.5
Consumption	36.8[e]	42.0 - 47.8	42 - 46
Oil Imports	8.2	7.5 - 11.3	4 - 5

[a] Range of recently published forecasts.

[b] Excluding exports of 0.8-1.0.

[c] Includes 0.1-0.3 liquid synthetics; 0.3-0.8 gas. Total synthetics from coal, oil shale, biomass are 1.0-1.5.

[d] Includes hydroelectric, solar, geothermal, biomass.

[e] Production plus imports do not equal consumption because of changes in stock levels and inclusion of product imports and NGL by volume, as commonly reported, rather than as crude oil equivalent.

reasonable and achievable. ESCOE, therefore, made no effort to identify further potential reductions, as this paper does.

Conclusion

In conclusion, several points about the 1990 U.S. energy situation should be re-emphasized:

- This paper speaks of a national goal of decreasing imports by perhaps as much as 50 percent by 1990. But there is nothing magic about 50 percent. The final result might be a reduction of 54 percent or 43 percent. No one -- no forecaster, planner, economist, engineer, geologist or geopolitician -- knows what 1990 will look like. This is evident not only from common sense but from the great variations among energy forecasts. So, energy decisions should be made with this uncertainty in mind. Whether or not the import reduction goal is actually met, it will be important in 1990 that the maximum effort had been made during the 1980s -- beginning as early as feasible.

- This paper cites ranges of possible 1990 production of the different energy sources in the accelerated case. Clearly there is nothing certain about that specific energy mix. Some planners will regard several of the accelerated case numbers as high and others as low. And, in 10 years, the 1990 scorecard may actually show somewhat less natural gas, somewhat more coal, some unexpected progress in synthetics. But _all_ resources must be developed, and continued conservation is essential. And, given the energy studies and forecasts reviewed in this paper, the ranges used for the accelerated case are within reach.

- While these ranges are reasonable, they will not be easy to attain. Enormous efforts will be required by the energy industry to reach even some of the lower production levels of the 1990 accelerated case. In the fields of oil and natural gas, for example, domestic production peaked in the early 1970s and has been declining ever since (with the oil decline slowed by the advent of Alaskan North Slope oil). Even in the accelerated case, it is unlikely that enough new discoveries will be made to sustain current production rates. But with the right decisions, production probably can be stabilized at least _near_ current levels.

- The accelerated case is consciously conservative on 1990 energy use. It assumes a high of some 46 million barrels a day of oil equivalent consumed in 1990. While this is a much smaller figure than forecasters were projecting for 1990 only a few years ago, it is still 3 million to 4 million barrels a day higher than some current oil company forecasts for 1990. Lower consumption in 1990 than assumed in this paper would make imports lower.

- Finally, this paper does not set a goal and then calculate how to reach it. Rather, it examines the resource base, studies various forecasts to derive consensus figures, analyzes current government policies which inhibit domestic production, cites general areas of policy change and, on a foundation built of those ingredients, describes what the overall U.S. energy situation could be by 1990 and what must be done to get there.

Our country must recognize that our heavy dependence on imported oil is not inevitable and move now to lessen that dependence significantly.

Clearly, movement toward reduced dependence on imports would provide significant benefits for Americans as individuals and for this nation as a whole.

As a nation, America would gain a stronger dollar, improved balance of payments, strength to inhibit political blackmail from abroad and more security against disruption of our vital oil supplies.

As individuals, Americans would gain more goods and services, more secure jobs, less inflation and higher wages and salaries.

II. NATIONAL SECURITY

The modern economic, military and international political worlds depend heavily on energy, and the most pervasive and flexible source of energy is oil. Thus it is reasonable to assess the possible impact on United States national security of an oil import reduction of as much as 50 percent.

National security involves far more than our country's war-fighting capability. It includes our ability to deter an attack, the military strength of our allies, our relationships with those allies, their ability to deter attacks, the flexibility of our foreign policy, the economic viability of our allies and our own economic strength.

The Authorities Agree

Authorities in and out of government agree that more secure energy supplies would improve both our economic and military security.

President Carter

President Carter has stressed the military/economic linkage for national security:

> Our country is at one end of a 12,000-mile supply line and half the oil that we use is on the other end of that supply line. Energy security is a vital link also between national military security on the one hand and economic security at home.[1]

The Congress

In August 1977, Congress reaffirmed the dangers of our dependence -- which was heavy then and is heavier now. In enacting the Department of Energy Organization Act, Congress declared that the "energy shortage and our increasing dependence on foreign energy supplies present a serious threat to the national security of the United States."[2]

This echoed congressional findings as long ago as 1973 that "oil shortages will create severe economic dislocations and hardships which constitute a national energy crisis threatening the public health, safety and welfare."[3]

Department of the Treasury

A Department of the Treasury report last year declared unequivocally that "the monetary repercussions accompanying the growing dependency on imported oil constitute a threat to the national security."[4]

In a memorandum to the president dated March 14, 1979, then-Secretary of the Treasury W. Michael Blumenthal went back several years in discussing the energy threat to national security. He wrote:

> On January 14, 1975 . . . Treasury Secretary [William] Simon found that the nation's dependence on imported oil was so great as to threaten to impair the national security and recommended to President Ford that action be taken to remove the threat. That conclusion is, unfortunately, even more valid today.
>
> The nation's dependence on imported oil has increased dramatically since the 1975 finding. At the time of Secretary Simon's finding, 37 percent of United States demand for oil was supplied from foreign sources. In 1978, oil imports accounted for 45 percent of oil consumed in the United States. During that same period, the nation became more dependent on oil to meet overall energy demand, and oil imports increasingly originated in a small number of distant foreign countries. The increasing dependence on foreign sources of oil is a consequence of both rising levels of consumption and declining domestic production.
>
> This growing reliance on oil imports has important consequences for the nation's defense and economic welfare. Because so much of the oil used in the United States originates thousands of miles away, supplies are vulnerable to interruption from a variety of causes. Recent developments in Iran have dramatized the consequences of this excessive dependence on foreign sources of petroleum. Furthermore, the rising level of oil imports adversely affects our balance of trade and our efforts to strengthen the dollar; in 1978, outflows of dollars for our oil imports amounted to $42 billion, $15 billion more

than in 1975 and offsetting much of the rise in our exports of industrial and farm products.

The continuing threat to the national security which our investigation has identified requires that we take vigorous action at this time to reduce consumption and increase domestic production of oil and other sources of energy. To the extent feasible without impairing other national objectives, we must encourage additional domestic production of oil and other sources of energy, and the efficient use of our energy supplies, by providing appropriate incentives and eliminating programs and regulations which inhibit the achievement of these important goals.[5]

Edward Erickson

On the strategic military side, a comprehensive study of the importance of oil to the United States has been undertaken by Professor Edward W. Erickson of North Carolina State University. Writing in the July-August 1978 issue of Current History, Dr. Erickson made these basic points:

- The United States is one of the superpowers upon which the balance of world military and political forces rests.

- The United States depends for approximately one-half of its oil consumption upon imported supplies.

- Increased U.S. oil imports have increasingly originated from potentially unstable Eastern Hemisphere sources of supply. (Twenty-five percent of U.S. oil imports, 60 percent of West European oil imports and 70 percent of Japanese oil imports come from the countries that border the Persian Gulf. Furthermore, the lengthy supply routes and several "choke points" through which critical shipments of oil to the industrialized nations must pass are vulnerable to direct military attack.)

- Western Europe and Japan are parts of a common economic structure and military defense system with the United States.

- Even were the United States completely energy self-sufficient, it would be strategically vulnerable because of its intertwined interests with

Japan and Western Europe -- which are more heavily dependent upon imported oil than is the United States.

- . . . All of the tactical forces [of the United States and its allies] are fueled by oil.

- The global interests of the superpowers in the contested areas of the world are attempted to be advanced, either directly or by proxy combatants, through limited conventional warfare -- and limited conventional warfare depends upon oil for air power, mechanized armor and logistic support.

- In the absence of severe nuclear proliferation and failures of restraint, conflicts between smaller countries which may be originally unconnected to rival superpower contentions will be one form or another of conventional warfare and hence, to varying degrees, oil dependent.

- The Soviet Union is the largest oil producer in the world and is essentially energy self-sufficient.

- Although there is dispute concerning the extent of the Soviet oil and [natural] gas resource base and the speed with which it can be developed, it is at least possible and perhaps likely that the Soviet Union could become a major source of energy supply for Western Europe.[6]

The Department of Defense

And the Defense Department has said:

The concentration of oil production facilities in the [Persian Gulf] area presents the major physical risk. This creates a risk of interdiction, or even the risk of natural or accidental disturbances. The extensive damage in the Abqayq fires in May and June 1977, caused by accident, highlights the fragility of these facilities. Destruction of key facilities could cause major interruptions of oil deliveries in the U.S. and to our NATO and Japanese allies which would adversely affect U.S. and Western World political, economic and military security.[7]

In a June 20, 1980, speech, Defense Secretary Harold Brown added:

In the last analysis, it is the ability of the United States and its allies to maintain a credible deterrent to Soviet aggression which underwrites not only our own security, but that of most of the oil-producing countries as well.[8]

The Central Intelligence Agency

Similarly, Stansfield Turner, director of the Central Intelligence Agency, told the Senate's Committee on Energy and Natural Resources:

> Moscow no doubt will make an intense effort to obtain oil at concessionary prices from the oil-producing countries through barter deals, sometimes involving arms sales. More forceful action, ranging from covert subversion to intimidation, or, in the extreme, military action, cannot be ruled out.[9]

The Central Intelligence Agency has said this of possible supply interruptions in producing nations, emphasizing price pressures:

> There is a high probability that acts of nature, human error or deliberately targeted terrorist attack will interrupt the flow of oil in one or more of the oil exporting nations during the next several years.
>
> Interruptions of oil supply owing to guerrilla operations, acts of terrorism, or acts of nature are not likely, by themselves, to be of a magnitude and duration which would result in severe economic disruption of Free World economies, though they could exert strong upward pressure on prices in a tight world oil market. Extensive terrorist action against key oil storage and transportation facilities in the Persian Gulf could, in particular, significantly affect the market by substantially reducing oil supplies for the time required to put those facilities back into operation, which could be several months.[10]

Military Oil Requirements

While wars have been fought over energy, and interruptions in supply potentially limit the ability of a country to fight for any reason, it should be emphasized that the potential dangers from our oil over-dependency are not from a present or future inability to fuel tactical fighter planes or tanks.

In peacetime, the needs of the U.S. military establishment for imported or domestic oil are small relative to U.S. supply. According to the Department of Defense, fiscal 1978-79 military requirements for oil amounted to 459,000 barrels a day -- or about 2.5 percent of total U.S. oil consumption during that period.[11]

This country clearly can meet all of its peacetime military petroleum needs from domestic oil production and storage. While the U.S. strategic petroleum reserve now holds slightly more than 90 million barrels of crude oil, it is eventually expected to hold 1 billion barrels.

We could have more difficulty adapting to a lengthy interruption in imported supplies as our stored reserves declined, but the nation could adjust, with some hardship, by instituting severe measures such as energy rationing. We could also hope to rely on the resources of nearby nations such as Canada, Mexico and Venezuela.

Economic disorder from a prolonged energy shortfall would hamper production of necessary military equipment and supplies and social disruption could cause severe recruitment and morale problems within the armed forces.

Vulnerability of Allies

Finally, the importance of our allies to our national security should be re-emphasized.

Klaus Knorr, professor of public and international affairs at the Woodrow Wilson School, Princeton University, comments:

> The United States cannot be indifferent to the even greater vulnerability of its allies and main trading partners, particularly the Western European nations and Japan. Any serious disruption of their economies threatens vital foreign policy interests of this country.[12]

And, the oil-consuming nations may be forced to compete for limited supplies. Such competition can be divisive and potentially threatening to political harmony among allies.

Of course, the United States cannot solve the energy difficulties that confront our allies. But reduction of our own dependency by perhaps as much as 50 percent can help them. Our oil demand has contributed greatly to tight world markets in which a relatively small cutoff can produce widespread dislocation. An easing of that demand would help lessen the effect of cutbacks and remove some of OPEC's leverage. With its rich energy resource base, the United States has the opportunity to reduce our vulnerability and that of our energy-poorer allies.

Conclusion

In sum, the fundamental truths of the government position and of Professor Erickson's basic points are unchallenged: United States flexibility to conduct international relations and to create an appropriate foreign policy is reduced when we live under threat of sudden disruption of a significant portion of a product as critical to our security as oil.

Several of the main oil-exporting countries are in the politically volatile Persian Gulf area. Hence, the United States and the other free world nations are particularly vulnerable to a cutoff of supplies. A government that is hostile toward us may simply decide to stop sending us oil. Also, an oil-exporting country may unexpectedly be thrown into turmoil by internal conflict or an attack from outside and thus be *forced* to stop sending us oil.

But even without a severe disruption in supply, the reliance of the United States on so much imported oil causes grave national security problems.

An import bill of the estimated $90 billion for 1980 represents an unprecedented transfer of American wealth to foreign nations, giving them increasing control over United States economic affairs.

Our continued dependence on significant supplies of foreign oil makes us particularly vulnerable to large, sudden price increases. To the extent that the OPEC cartel can maintain cohesive policies, it can dictate -- within limits -- both the price and the availability of the oil on which the free world depends. The cartel demonstrated this control during the 1973-74 embargo, when it quadrupled the price of OPEC oil exports within five months. And in 1979 a relatively small supply reduction again tightened the world market enough to allow a huge price increase. Iran's new government cut its production by 2 million barrels a day -- only 4 percent of free world supplies -- and OPEC doubled prices.

All of these problems, and particularly those of our allies, do not disappear if United States imports are reduced to 4 million or 5 million barrels a day. But, clearly, reduced dependence would give us additional options and flexibility which would enhance our national security. If we continue on our present course, choosing the 1990 world to which that course is leading, the security options will be fewer and the flexibility less.

III. OIL AND NATURAL GAS

Some analysts believe that more oil and natural gas will be found in America in future years than in all of the past. This country is a long way from exhausting its potential supplies of these two fuels. And all experts agree that these will be principal sources of energy in the United States well beyond 1990.

The debatable question, in considering the national energy mix for 1990, is precisely how much oil and natural gas will be produced domestically. For many years, the United States has been using considerably more domestic oil and natural gas than it has been finding. As a result, production has been declining. Will this situation be stabilized or will the decline inevitably continue?

No one knows precisely what oil and gas production levels will be in 1990; no one can know without additional exploration and drilling. Thus, the nation should not plan on the basis of the most optimistic forecasts or the most pessimistic but should increase domestic production as much as is feasible and should eliminate unnecessary delay in determining what our resources are.

As discussed earlier, many 1990 forecasts were studied for this paper, and extractions from them offer a composite view of 1990 energy production.

Production of oil and natural gas liquids now averages about 10 million barrels a day. The forecasts reviewed for this paper (see page 13) show a 1990 range of 6.1 million to 10.4 million barrels daily.

U.S. natural gas production last year was about 19 trillion cubic feet -- the energy equivalent of 9.1 million barrels a day of crude oil. The forecasts studied (see page 14) show a range of 12.5 trillion to 20.9 trillion cubic feet a year (the equivalent of 6.1 million to 10.2 million barrels a day of oil) by 1990.

Thus, for both oil and natural gas, the forecasts see a sizable potential decline in production over the coming decade.

But those projections are based on what is likely to happen, not on what inevitably will happen.

If we as a nation choose to make efficient use of our potential petroleum resources, if the public and government choose to move in the direction of cutting 1990 imports by as much as 50 percent, within the decade we can stabilize oil and natural gas production near today's levels. As indicated in Chapter I, the accelerated case suggests the possible 1990 production of some 9 million to 10 million barrels a day of crude oil and natural gas liquids, plus or minus, and some 16.5 trillion to 19 trillion cubic feet of natural gas a year, plus or minus (the equivalent of about 8 million to 9 million barrels a day of oil).

Two points need emphasis:

1. An enormous industry effort will be required to reach even the lowest projected figures. Declines have taken place in the production of both crude oil and natural gas -- as older fields continue their natural decline and new discoveries fail to replace supplies being used up. Recent exploratory results in such areas as the Gulf of Alaska and the Baltimore Canyon off the New Jersey coast have not been encouraging, but experts believe that large new domestic discoveries of oil and gas are still likely to be made in the western states, in Alaska and on the Outer Continental Shelf (OCS).

Under present conditions, lead times between initial exploration and first production in some frontier areas are estimated to be 8 to 11 years. Thus, rapid exploration and development of the oil and gas resource potential of the OCS and other areas are essential to help offset declining production. The rate of growth in oil and gas production will depend largely on governmental leasing and regulatory policies. In the accelerated case estimates of this paper, it is assumed that those government policies will be improved.

2. The national objective proposed in this paper is to make this nation as energy secure as possible by 1990. To do this, the effort must be made now to find and produce more oil and natural gas. If, after the maximum effort has been made, less oil and natural gas are found and produced than any forecasts suggest, we can pursue other routes available to us. The import reduction goal may still be met by additional production from other sources or less consumption than now projected. In any case, the nation will be better off if the effort is made now to find more oil and natural gas.

Six basic facts about oil and gas will be addressed in this chapter:

1. More domestic oil and gas can be found and produced.

2. Major future discovery areas are the western states, Alaska and offshore.

3. Government controls access to promising areas.

4. Government affects petroleum development through environmental laws and regulations.

5. Decontrol and higher prices increase investment in exploration and production.

6. Oil companies are working to find more petroleum.

A summary section will review some of the positive steps that the federal government can take to assure careful and efficient use of the natural resources contained in the federal lands onshore and offshore.

A supplement will list several specific ways in which oil companies are improving this country's energy situation.

* * * * * * *

1. <u>More Domestic Oil and Gas Can Be Found and Produced</u>.

Proved reserves of crude oil and natural gas in the United States have declined every year since 1970, when statistics first reflected the oil and gas discoveries at Prudhoe Bay on Alaska's North Slope.

Between 1970 and the end of 1979 this country's proved oil reserves dropped more than 11.9 billion barrels, while natural gas proved reserves declined 95.8 trillion cubic feet -- the energy equivalent of 17 billion barrels of oil.

In the eight-year period 1971-1978 this country used up its proved oil reserves almost twice as fast as new supplies were being found. During those years, the United States withdrew a yearly average of 3 billion barrels of oil from proved reserves while replacing them with only 1.6 billion barrels of newly found petroleum.

Those trends obviously placed a serious drain on the nation's "bank account" of proved reserves. When proved reserves shrink, the nation's ability to produce oil and gas from those reserves naturally follows the same downward path.

As a result, domestic production of oil and natural gas liquids peaked at 11.3 million barrels a day in 1970 and declined to about 10 million barrels a day in 1979. Natural gas production peaked at about 11 million barrels a day of oil equivalent in 1973 and declined to about 9 million barrels a day equivalent in 1979.

The domestic oil decline has been slowed by the advent of Alaskan North Slope oil, which began flowing through the trans-Alaska oil pipeline in 1977. North Slope production rose to 1.5 million barrels a day in late 1979. However, for the nation as a whole, the trend is still downward.

But those declines need not continue throughout the decade.

The latest estimates, published in May 1980 by the American Petroleum Institute and the American Gas Association, indicate that 1979 may have marked a turning point. Although proved reserves of both oil and natural gas again declined, the rate of decline slowed down. Additions to oil and gas reserves equaled approximately three-fourths of domestic production. As a result, the net decreases in oil and gas reserves were the smallest since the Prudhoe Bay discovery.[1]

As 1979 ended, the nation's proved reserves included:

- 27.1 billion barrels of crude oil;
- 5.7 billion barrels of natural gas liquids; and
- 194.9 trillion cubic feet of natural gas (the equivalent of 34.7 billion barrels of oil).

The API report also estimated that an additional 3.6 billion barrels of oil could be added to proved reserves if and when improved recovery techniques are successfully applied to known fields.

The additions to proved reserves during 1979 reflected sharply accelerated drilling activity and rapid expansion of efforts to improve recovery in existing fields. The beginning of phased decontrol of oil prices and an increase in the controlled price of natural gas both played an important role in making these activities possible.

Proved reserves estimates represent supplies which have already been found by drilling. These reserves are believed to be recoverable in future years under economic and operating conditions as they now exist.

These figures, however, do not represent this country's total potential ability to continue finding and producing oil and natural gas.

This country can continue finding and producing oil and gas by:

- exploring onshore and offshore frontier areas which have never been tested with the drill;
- drilling deeper in areas which are already productive;
- reexamining older, mature producing areas such as the states of Pennsylvania, New York, Ohio and West Virginia;
- using more sophisticated technology to find petroleum; and
- applying the most efficient recovery methods available to coax as much oil and gas as possible out of the earth. There are 300 billion barrels of known oil resources in the United States which until now have been uneconomic to produce. If even 10 percent of that oil can be produced through enhanced recovery methods now being developed, today's proved oil reserves would double.

Prophets of Gloom Were Wrong

Over the past century there have been many gloomy predictions about this country's petroleum potential:

- In the 1880s and 1890s a government bureau said there was little or no chance of finding oil in California, Kansas or Texas. Today, these are among the leading oil-producing states.

- A 1920 government report said the United States had almost reached its peak in oil production, which was then 443 million barrels a year. In later years the nation's oil output climbed to more than 3.5 billion barrels a year before declining to current levels of around 3 billion barrels annually.

- In 1939 a government official said America's oil supplies would last only 13 years.

- In 1949 a federal official announced that the end of United States oil supplies was almost in sight.

• Until five years ago, many people said oil and gas would never be found in the Overthrust Belt of the Rocky Mountains. But oil and gas now have been found there, and it is one of the most actively explored areas in the United States.

There have always been discouraging predictions. But drillers have kept on hoping and working. As a result, today 33 of our 50 states produce oil, natural gas, or both.

Analysts Agree: Potential Is Great

Studies by government agencies, scientific organizations, universities and oil companies have concluded that this country has the potential to produce large amounts of oil and gas for many years to come.

Those studies of the undiscovered recoverable oil and gas resources of this country were made by different persons at various times during the past few years. They used varying geologic data, production histories and assumptions concerning economics and technology. So naturally they reached somewhat different conclusions.

Dr. Charles D. Masters, chief of the Office of Energy Resources in the U.S. Geological Survey (USGS), reviewed several such studies in a speech in January 1979. He said that although the estimates vary in detail, they indicate that at least for the next 30 to 50 years, resource potential will not be a limiting factor in sustaining U.S. oil and gas production.

The limiting factor, Dr. Masters said, will be the rate at which this country can find those supplies and convert them from potential to proved reserves.[2]

How much of this oil and gas will be found and produced -- and how soon -- will depend on many factors, including pricing, incentives for investment, technological progress, access to promising new areas to explore, environmental restrictions and the overall political and economic climate.

We Need to Move Ahead Rapidly

The differences of opinion reflected in the varying estimates of this country's oil and natural gas potential serve a useful purpose. They emphasize the need to accelerate leasing and drilling of promising prospects so that the nation can base its future plans on as many facts as possible.

Even under the best of circumstances, long lead times may be required to find significant new supplies of oil and natural gas, develop them, and create facilities to process them and transport them to the ultimate consumers. The downward trends in reserves and production, which have been developing for the past decade or more, cannot be reversed overnight. In some frontier areas -- such as Alaska, the Rocky Mountains or the Outer Continental Shelf -- several years may be needed to deliver new supplies to market, even after the first discovery wells are drilled.

That fact underscores the need to eliminate all unnecessary delays and to move forward rapidly to find out, as precisely as possible, how much domestic oil and gas this country can depend upon.

How Much Can We Produce?

The USGS is the source of the most widely quoted estimates of the amounts of oil and gas this country can expect to produce in the future.

In 1975 the USGS published a range of estimates of oil and gas yet to be discovered but believed to be economically recoverable. The same report contained data on proved reserves and estimates of petroleum that could be recovered from known reservoirs by injecting water or other fluids to increase oil production. The federal agency's estimates indicated that the nation's future production of crude oil and natural gas liquids could be expected to range between 135 billion and 223 billion barrels, with a statistical mean of 172 billion barrels.

For natural gas, the production estimates ranged between 761 trillion and 1,094 trillion cubic feet (tcf), with a mean of 923 tcf (the energy equivalent of 164 billion barrels of oil).[3]

In March 1980 the USGS revised the amounts of oil and gas estimated to be recoverable in certain offshore areas, but did not change the national total significantly.[4] The agency is preparing a new national study which may be published later this year.

Pending completion of that study, any current estimate of future oil and gas production -- onshore and offshore -- has to take into account the declines in proved reserves since 1975. Therefore, industry spokesmen have told congressional hearings that a reasonable estimate of the nation's remaining economically recoverable petroleum is:

- 150 billion barrels of crude oil and natural gas liquids; and

- 800 trillion cubic feet of natural gas (the equivalent of 142 billion barrels of oil).

To put those figures into perspective, that would be:

- more than 40 times the current annual production of oil and gas in the United States; and
- more oil and gas than this country has produced in the 120-year history of the petroleum industry.

But this is only part of the story. The USGS estimates of economically recoverable oil and gas do not include oil and gas that are now uneconomic but might possibly be recovered with higher ratios of price to cost and improved recovery technology. That category includes an estimated 165 billion to 250 billion barrels of oil and 130 trillion to 200 trillion cubic feet of natural gas (the equivalent of 23 billion to 35.6 billion barrels of oil). The USGS says between 70 percent and 75 percent of this sub-economic petroleum already has been identified, but ways have not been found to produce it economically.

The Geological Survey's estimates of economically recoverable oil and gas also do not include energy that may be produced from oil shale, tar sands, heavy oils, gas locked in tight rock or coal formations, or offshore oil and gas in water depths beyond 200 meters.

The USGS points out that when these sources become economic they will expand the future potential by tens or hundreds of billions of barrels of oil and trillions of cubic feet of natural gas.

On May 16, 1980, the USGS published its first estimates of the hydrocarbon potential of deeper Atlantic waters beyond the Outer Continental Shelf. Using a range of estimates, the agency indicated that a 1.4-million-acre portion of the continental slope off New Jersey and Maryland may contain as much as 7.3 billion barrels of oil and 28.5 trillion cubic feet of natural gas (the energy equivalent of 5.1 billion barrels of oil).[5] The Interior Department has tentatively scheduled a lease sale in that area for December 1981.

2. Major Future Discovery Areas Are the Western States, Alaska and Offshore.

The vast majority of the untested areas believed to have oil and gas potential are controlled by the federal government, which has not made the great bulk of those lands available for exploration and drilling.

120 Years of Production

This country has been producing petroleum for 120 years. The nation's first oil well was completed at Titusville, Pennsylvania, in 1859. It found oil at a depth of only 69 feet.

Since then about 2.5 million wells have been drilled in the United States in search of oil and gas. In recent years some wells have probed as deep as 30,000 feet -- more than 5 miles.

During most of the past century there were many geologically attractive areas in private ownership. Oilmen could negotiate leases with landowners and start drilling. Landowners received a share of the value of any oil or gas produced.

Today, about 20 percent of the total land area of the United States is either producing oil and gas or is under lease for exploration. About three-fourths of the land under lease is privately owned.

Where the Best Chances Are

Still largely unexplored and untested, however, are hundreds of millions of potentially productive acres whose resources are the common heritage of all Americans. These are the federal onshore lands, which are concentrated mainly in the western states and Alaska, and the federally controlled Outer Continental Shelf.

Studies by the USGS concur with analyses by industry and academic specialists that the nation's chances of finding major new oil and gas fields increase if leases are made available in these relatively untested areas.

The USGS has estimated that the continental shelf out to a water depth of 200 meters may contain 31 percent of the undiscovered oil and 21 percent of the undiscovered natural gas in the nation.

USGS studies also estimate that 37 percent of the nation's undiscovered oil and natural gas liquids and 43 percent of the undiscovered natural gas supplies lie on federal lands onshore and offshore.

Private companies pay substantial amounts to lease these lands. In 1979 alone, the federal government collected $4.6 billion in cash bonuses in five offshore lease sales, plus $1.5 billion in royalties based on offshore oil and gas production.

3. Government Controls Access to Promising Areas.

The federal government is, by far, the largest landowner in the United States -- onshore and offshore.

Onshore

The 50 states contain a total of 2.27 billion acres of land. The federal government owns 775 million acres -- one-third of the nation's total land area. These federal lands are located principally in the western states and Alaska. The government also retains control over the subsurface mineral rights on 63 million additional acres of onshore lands.

The onshore lands owned by the government are equal in size to the combined areas of 31 eastern states and the District of Columbia. That is the equivalent of all the land between the Atlantic Coast and the eastern borders of Texas, Oklahoma, Kansas, Nebraska and the Dakotas.

In many of the states regarded as having the highest oil and gas potential, the federal government is the largest landowner. Alaska is the prime example. As of September 30, 1978, the government owned more than 98.5 percent of all the land there. After completing pending land transfers to the state of Alaska and to Alaskan natives, federal agencies will still control about 62 percent of that state's area.

The federal government also owns more than 48 percent of Wyoming, 47 percent of California, 37 percent of Colorado and 33 percent of New Mexico. Some 70 federal departments, agencies, bureaus and other units own and administer federal lands. And other agencies -- such as the Environmental Protection Agency -- exercise considerable influence over the way federal lands are used.

Only 13 percent of the federally owned onshore lands -- 103 million out of 775 million acres -- are now leased for oil and gas operations. In 1979, wells on those leases provided 7 percent of the crude oil and natural gas liquids and 6.1 percent of the natural gas produced in this country. These percentages can be increased significantly during the 1980s through greater access to those lands.

Offshore

Newly revised estimates by the U.S. Geological Survey indicate that the nation's offshore areas out to a depth of 200 meters (including state and federal waters) may contain about one-fourth of the undiscovered recoverable oil and gas in the entire country.

USGS estimates of the undiscovered recoverable oil beneath our offshore waters out to a depth of 200 meters range between 12.5 billion and 38 billion barrels. The estimates for natural gas from the same areas range between 61.5 trillion and 139 trillion cubic feet (the equivalent of 10.9 billion to 24.7 billion barrels of oil).[6]

By comparison, proved reserves for the entire United States, onshore and offshore, at the beginning of 1980 were 27.1 billion barrels of oil and 194.9 trillion cubic feet of natural gas (the equivalent of 34.7 billion barrels of oil).

The federal government controls 528 million acres of submerged lands on the Outer Continental Shelf -- the area between the edge of state jurisdiction and the 200-meter water depth. (The limits of state jurisdiction are three miles from shore for most states and 10.5 miles for western Florida and Texas.)

From the beginning of federal offshore leasing in the 1950s until now, only 17.9 million of these 528 million acres have ever been leased -- less than 4 percent of the total. Many of the older leases have expired. At the end of 1979 only 10.3 million federal OCS acres -- 2 percent of the total -- were currently under lease for petroleum operations.

Yet those federal leases accounted for 9.1 percent of the crude oil and natural gas liquids and 23.5 percent of the marketed natural gas produced in the entire nation during 1979.

In older, more mature areas of the Gulf of Mexico, oil production is dropping rapidly. Oil output from federal leases off the Louisiana coast amounted to more than 358 million barrels in the peak year of 1971, but by 1979 had declined more than 109 million barrels.

While this kind of decline is normal in fields that have been producing for a number of years, it nonetheless underscores the need to find new productive areas to replace those being depleted.

An important new offshore field began oil production from federal leases in the Gulf of Mexico in September 1979. Identified as Mississippi Canyon Block 194, the field is estimated to contain 100 million barrels of ultimately recoverable oil.

Production is coming from under waters 1,025 feet deep -- the deepest yet tackled by the industry anywhere in the world.

Alaska

Two-thirds of the nation's continental shelf lies off Alaska's coasts. The USGS estimates that 70 percent of the nation's undiscovered offshore oil and more than 60 percent of the offshore natural gas may lie beneath Alaskan waters. The USGS has estimated that Alaska -- onshore and offshore -- may contain as much as 49 billion barrels of undiscovered recoverable crude oil and 132 trillion cubic feet of natural gas (the equivalent of 23.5 billion barrels of oil).

But while Alaska is estimated to have the greatest petroleum potential of any state, relatively few wells have been drilled there in search of oil and gas.

President Carter sent a message to Congress in January 1980 proposing that the National Petroleum Reserve in Alaska (NPRA) be opened, for the first time, to competitive exploration and production by privately owned companies.[7]

The NPRA is a 23-million-acre tract lying about 50 miles west of the giant Prudhoe Bay oil and gas field. This area has been held in reserve for nearly 60 years as a possible source of petroleum for the United States in time of national emergency. No private leasing has ever been permitted. From time to time the Departments of the Navy and Interior employed contractors to drill, but, on average, only one well has been drilled for each 822 square miles (526,000 acres) in the entire reserve area. (By contrast, 2.5 wells have been drilled for each square mile of land in Texas; in Oklahoma the ratio is 4.5 wells per square mile -- and drillers are still finding new oil and gas in those states.)

One oil field was found in the NPRA several years ago, but it has never been developed. A gas field supplies fuel for the village of Barrow, and another gas discovery was announced by the Interior Department on April 15, 1980. Obviously, if the nation is ever to rely on that remote area as a source of petroleum, a great deal more work must be done.

The president's plan calls for safeguarding the arctic environment, including the native culture and fish and wildlife resources of the North Slope. The president and the USGS director told Congress that development of NPRA's resources can be achieved more quickly and at less cost to the government through competitive operations by a number of companies than through any other approach.

In sum, much more oil and natural gas can be found and produced in this country if the federal government opens up more land -- onshore and offshore -- to petroleum leasing. Leases are now in effect on only 13 percent of the federally owned onshore lands and on only 2 percent of the federal offshore acreage. Yet, from those relatively small percentages of the total area under federal control, wells operating in 1979 accounted for:

- 16 percent of this country's oil and natural gas liquid production, and
- 30 percent of the nation's natural gas production.

Greater and more timely access to the vast unexplored lands owned or controlled by the federal government offers significant opportunities for the discovery of new reserves of oil and natural gas.

4. Government Affects Petroleum Development Through Environmental Laws and Regulations.

During the 1970s, the federal government adopted a series of major environmental laws in response to public concern about environmental pollution. These laws have brought about a substantial improvement in the nation's air and water over the past 10 years. The vigorous pursuit of environmental goals, however, has limited the development of this country's resources, including domestic oil and gas. It has also led to increased dependence on imported oil.

More Flexible Environmental Policies

To achieve the goal of cutting oil imports by as much as 50 percent by 1990, the nation needs more flexible environmental policies that permit both continued environmental improvement and increased domestic energy development. These objectives are not inconsistent.

Reasonable and revised environmental laws and regulations can:

- speed up development of the oil and natural gas on the U.S. Outer Continental Shelf;
- accelerate production of billions of barrels of heavy oil, primarily in California;

- increase recovery of oil in existing fields through use of expensive enhanced recovery techniques; and

- lead to exploration for new oil and gas supplies in areas of the nation that are presently off limits to petroleum development because of existing environmental laws.

Summarizing a recent study sponsored by the Ford Foundation, Philip K. Verleger, Jr., senior research scholar at the Yale School of Organization and Management, concluded that the "authors suggest that the present form of U.S. legislation is far more restrictive than is necessary to achieve the nation's environmental goals."[8] The study itself (Energy: The Next Twenty Years) characterizes the government's approach to controlling air pollution as "brute force regulation."[9]

Environmental Restraints and Roadblocks

Examples of petroleum-related projects cancelled or delayed because of overly rigid government constraints and excessive environmental opposition can be found from coast to coast. They include numerous instances of oil and gas production delays.

Some projects have been held up for years, causing loss or deferral of hundreds of thousands of barrels of oil and billions of cubic feet of natural gas. In southern Louisiana alone, a recent study found that production of more than 400,000 barrels of oil was either lost or deferred because of delays in the issuance of permits by the Corps of Engineers.

Oil companies estimated that the delays involving 55 permits during an eight-month period cost nearly $20 million. Thirty-two oil wells could have been drilled in southern Louisiana for that amount. The companies also estimated that 428,000 barrels of oil and 14.9 billion cubic feet of natural gas were unproduced because of the permit delays.

Facilities to transport and refine petroleum also are affected by opposition, restrictions and delays. A pipeline intended to carry Alaskan crude oil from California to Texas was cancelled after lengthy delays and changing conditions made the project economically unfeasible. A deepwater port off the coast of Texas also was cancelled.

The nation needs to modernize and rebuild much of its oil refining capacity in order to convert heavier grades of crude oil into the light products that are needed in today's market, to reduce the sulfur content of fuels, to make more unleaded gasoline and to improve the environment in other ways. Yet environmental opposition has played a major role in forcing the abandonment of many proposed refining projects in recent years, including several on the Atlantic Coast. More than nine years were spent in obtaining permits to build a refinery at Portsmouth, Virginia, and almost as much time trying to get permission to build a refinery at Eastport, Maine.

Existing constraints on energy development can be reviewed by Congress and remedial action can be taken to help the nation develop its energy resources while still protecting the environment.

(Further discussion of energy and the environment will be found in Chapter VIII.)

5. Decontrol and Higher Prices Increase Investment in Exploration and Production.

The gradual phasing out of federal controls on crude oil and higher prices allowed for natural gas are making it possible for U.S. petroleum companies to increase their investments in exploration and production.

Companies Are Investing More

In 1979, for example, 19 leading U.S. oil companies reported a total of $31.4 billion in worldwide capital and exploration expenditures. This represented 1.6 times their worldwide net income and 85 percent of their estimated cash flow.

Within the United States, those same 19 companies' 1979 capital and exploration expenditures totaled $21.3 billion. This was 11 percent more than their combined worldwide net income. Almost two-thirds of the U.S. total was for exploration and production. The 19 companies increased their 1979 expenditures for finding and producing petroleum by 44 percent over the level of the previous year.

Sixteen companies providing information for a recent survey said they were planning $37.5 billion in worldwide capital and exploration expenditures during 1980. This is an increase of 19 percent over their 1979 expenditures and is 1.8 times (almost double) their 1979 net income. Data for 11 of those companies indicate that 65 percent of the total planned capital and exploration expenditures are for projects within this country.

A study published in the February 18, 1980, issue of The Oil & Gas Journal said U.S. oil companies plan to spend 26 percent more on capital projects and exploration in this country in 1980 than they spent in 1979.[10]

The magazine said two-thirds of that money is targeted for U.S. drilling, exploration, production and buying offshore leases. The other one-third will pay for improving and expanding facilities for petrochemicals and for refining, transporting and marketing petroleum products. The companies are upgrading refineries to handle heavier crude oils and to meet environmental requirements. They also are building more oil and gas pipelines, modernizing tanker fleets and creating more storage capacity.[11]

Operating Costs Are High

An important factor in increased expenditures is the high cost of drilling. In 1968 it cost an average of more than $61,000 to drill an onshore well in the United States. By 1978 it cost more than $230,000 -- about 275 percent more.

In the same period the cost of the average offshore well rose from about $505,000 to more than $2.1 million, an increase of about 320 percent.

These are only average costs. Deep wells, onshore or offshore, can cost many millions of dollars. For example, the cost of a 23,254-foot onshore well drilled in Mississippi early this year was estimated at $42 million.

A Department of Energy (DOE) study found that the installed costs of surface production and lease equipment increased from 1970 to 1977 by an average of more than 100 percent for new land-based operations in all major producing areas of the lower 48 states. DOE said the increase in the cost of installed equipment was much higher than the annual inflation rate of all goods and services in the United States. Operating costs, according to the DOE report, increased by as much as 155 percent in some areas during the same period.[12]

The federal government can adopt economic policies which give inherently efficient market forces a better chance to direct a stepped-up search for domestic oil and gas. The economic system needs efficiency and flexibility to take advantage quickly of developing energy opportunities. The success of this country's efforts to increase domestic oil and gas production will depend heavily upon the oil companies' ability to attract the level of investment capital that will be needed. Federal government policies will have a major impact on that ability. Key policies needed are the continuing phase-out of price controls on oil, the removal of controls over natural gas, elimination of artificial governmental restrictions on the distribution of petroleum and adoption of sensible tax policies that encourage exploration and production.

6. Oil Companies Are Working to Find More Petroleum.

The nation's oil companies are searching vigorously for the oil and gas supplies this country needs. They have the technology and the skilled personnel to operate with a high degree of efficiency. From Alaska to the Gulf of Mexico they have proved that petroleum operations can be carried on in harmony with the nation's environmental goals.

Improved prices for domestic producers are helping the oil companies to:

- speed up the search for new oil and gas fields;
- go after oil and gas which could not have been produced economically under the old federal price ceilings;
- use advanced technology to increase oil recovery from existing fields; and
- keep many older wells active longer.

Some 12,000 large and small producers are now trying to find and produce oil and gas in this country. They have increased their drilling efforts sharply in the past decade.

In the years 1970-1979, oilmen drilled 358,407 wells in the United States. That represented about 80 percent of all the wells drilled in the entire free world.

Companies drilled 49,816 wells in the United States in 1979 -- more than in any 12-month period in the past 20 years. This was an increase of nearly 93 percent over the low point reached in 1971 and reflected a movement toward the 1956 record of 57,077.

In the first quarter of 1980, companies completed 13,217 wells -- an increase of nearly 13 percent over the first quarter of 1979.

During the first 24 weeks of 1980, the Hughes Tool Company reported the average number of rotary drilling rigs at work as 2,686. That figure exactly matched the annual record set in 1955. It also represented an increase of 32.5 percent over the count for the comparable weeks of 1979. On June 16, 1980, the Hughes report showed 2,857 rigs at work. This was an increase of 877 rigs, or 44 percent, over the single-week figure reported in mid-June 1979.[13]

Two current examples document the quickening pace of domestic exploration and drilling:

- One oil company alone recently announced that it had spent nearly $100 million drilling in the Overthrust Belt area of the Rocky Mountains in 1979, and plans to spend more than $130 million in that area during 1980.

- The California State Division of Oil and Gas reported that, during the first two months of 1980, drilling activity in that state increased nearly 60 percent over the same period a year earlier.

Reexamining Old Producing Areas

It is an old saying -- but a true one -- that a good place to search for petroleum is where it has been found in the past. Producers today are taking a new look at many areas that were productive in the past. They are also exploring areas which did not seem promising with the technology that was available several decades ago, but which may be more accessible today.

Among the areas being reexamined are some of the pioneer producing states where the U.S. petroleum industry was born more than a century ago -- Pennsylvania, New York, Ohio and West Virginia. In the three-year period 1977-1979, oilmen drilled 18,999 wells in those four states: 8,687 in Ohio, 5,038 in Pennsylvania, 4,003 in West Virginia and 1,271 in New York.

The latest proved reserves estimates show that the renewed emphasis on that region is producing results. During 1979 all four of those states increased their proved natural gas reserves, and all but West Virginia showed a gain in proved crude oil reserves. For the four-state area, oil reserves increased 8.1 million barrels in 1979 as compared to 1978, while gas reserves increased more than 510 billion cubic feet (the equivalent of nearly 91 million barrels of oil). Although no giant fields have been found in those states, every producing well helps to meet the nation's energy needs and to move toward a goal of cutting oil imports in half by 1990.

Other states along the Appalachian Mountain chain also are being studied for their oil and gas potential, using today's sophisticated seismic techniques and computerized analyses of the seismic data.

Drill Bits Probe Deeper

In other parts of the nation, oilmen are drilling deeper in areas which have been productive for years. They are trying to find supplies which may have been missed in earlier, shallower drilling. Three such areas are Louisiana, the Williston Basin of North Dakota and Montana, and the Anadarko Basin of Texas and Oklahoma.

- In both onshore and offshore Louisiana there has been a great deal of deep drilling for gas within the past five years. New types of seismic equipment are helping producers to map those deep areas more accurately. At least four important gas fields have been found, and the search is continuing.

- The Williston Basin was discovered in the 1950s and has been an important oil-producing area ever since. As is always the case in a mature or aging area, production has been declining in recent years. However, recent drilling to depths of 12,000 feet or below has resulted in several discoveries.

- The Anadarko Basin was found in the 1920s. Recently, deep drilling for gas has increased in this area. The Wall Street Journal reported on March 27, 1980, that companies drilled 140 wells of more than 15,000 feet in the Anadarko Basin in 1979. This was almost twice as many deep wells as were drilled in that area the previous year.

Perseverance Pays Off

Drilling costs are extremely high at great depths. Typically, it can cost $1 million to drill the first 12,000 feet; another $1 million to reach 15,000 feet; and $2 million more to reach 18,000 feet.

The Wall Street Journal of March 27, 1980, also pointed out that companies are drilling more wells at a higher cost per well. According to the newspaper, companies spent more than $2 billion in 1979 to drill 614 wells of 15,000 feet or more in various parts of the country. Two years earlier they spent about $1 billion to drill 414 such wells. Based on those figures, the average cost per deep well increased by 33 percent in two years -- from $2.4 million per well in 1977 to more than $3.2 million each in 1979.[14]

Experience has taught oilmen that searching for oil and gas thousands of feet below the earth's surface is not an exact science. For instance, they drilled 15 dry holes in northern Alaska over a 10-year period at a total cost of $500 million before finding the giant Prudhoe Bay field in 1968.

Perseverance and experience are producing results in other areas as well. For example, for many years the Overthrust Belt of the Rocky Mountain states was regarded by oilmen as a "driller's graveyard." It is an area where natural forces thrust old, extremely hard rocks over younger, softer sedimentary rocks that might contain oil or gas. A similar situation exists in the Appalachian region of the eastern United States.

More than 300 dry holes were drilled in the Rocky Mountain Overthrust area before the first commercial discovery about five years ago. But now at least 12 commercial oil and gas fields have been found in that area, proving that it has great potential. Here, again, improved technology is helping drillers find petroleum.

Oregon became the thirty-third producing state in 1979 after the discovery of commercial quantities of natural gas. Drillers had searched for oil and gas in that state for 78 years. The discovery has touched off a wave of exploration in Oregon, with 3.5 million acres currently under lease.

Producing More Oil from Known Fields

Although exploratory drilling will be extremely important in maintaining petroleum production, much of the oil this country will need in coming years does not have to be discovered. Existing fields contain an estimated 300 billion barrels of oil which until now have been economically unproducible.

Continued improvements in economic incentives and technology, along with reasonable environmental regulations, will make it possible to recover more of that oil.

Conventional production methods, on average, have recovered from the ground only about one-third of all the oil ever discovered in this country. Forty years ago oilmen developed a technique called waterflooding, in which water is injected underground to force oil to move toward producing wells. Today about half of the oil produced in the United States is the result of waterflooding, but that method still leaves large amounts of oil in the ground.

In recent years oil companies have developed a variety of newer methods to increase oil recovery. They involve the use of heat, chemicals and various gases and fluids to thin out oil in rock formations, loosen it from the tiny pores in the rocks or otherwise cause it to flow toward producing wells. These methods are expensive to use and some of them require months or years to produce significant results. When much domestic oil production was held below a price ceiling of about $5 per barrel, such methods were far too costly to justify their use.

Under current laws and regulations, oil produced through the use of certain recovery techniques is exempt from federal price controls. The gradual decontrol program also provides additional incentives to producers who invest in enhanced recovery projects. Oil produced with these new methods is subject to the "windfall profits" tax, but at a lower rate than most domestic production.

With adequate incentives and continued technological advances, most estimates indicate that future production of oil in the United States could be increased by around 30 billion barrels with the use of the newer techniques. That would represent a doubling of current proved reserves.

A survey published in the March 31, 1980, issue of <u>The Oil & Gas Journal</u> concluded that enhanced recovery methods are currently adding about 385,000 barrels a day to the nation's oil production. That represents about 4.5 percent of total daily output, with more than three-fourths of the enhanced recovery being achieved through the injection of steam. The survey says the second leading type of treatment involves the injection of carbon dioxide and other gases.[15]

The magazine listed a number of enhanced recovery projects for which oil companies and the Department of Energy are providing joint financing. Private companies are providing more than $132 million for those projects and the government is adding nearly $94 million, for a total of $226 million.[16]

A recent DOE study projected that the nation's production of oil through enhanced recovery methods could increase to between 1.3 million and 1.4 million barrels a day by 1990 and even more by 1995. The government study pointed out, however, that the number of enhanced recovery projects undertaken, and the success of those experiments, will depend largely upon pricing, return on investment, technological advances, environmental restrictions and the availability of manpower and materials. The DOE emphasized that many of the methods discussed are still in the initial stages of development and that long and costly efforts will be necessary to bring them to successful commercial application.[17]

Companies' Role in Research

U.S. petroleum companies concentrate most of their research and development (R&D) efforts on developing improved technology for finding, producing, transporting and refining petroleum. But, along with other leading U.S. industries, oil companies also are working to develop new forms of energy.

Oil companies have developed highly sophisticated equipment for recording, processing and analyzing seismic data, which are vital in deciding where to drill.

They have also developed production platforms capable of operating in water depths far beyond the reach of conventional offshore equipment; built computerized models to simulate the production response of underground oil reservoirs; invented new catalysts to increase gasoline yield at refineries by as much as 40 percent; and developed safer transportation methods for oil and petroleum products.

National Science Foundation data show that oil companies carry out 45 percent of all energy R&D funded by private companies in the United States.[18] Records in the U.S. Patent and Trademark Office show that 18 oil companies accounted for more than 40 percent of the R&D patents awarded in seven out of 10 categories of synthetic fuels during the decade 1969-1978.[19]

A study published by INFORM, Inc., a non-profit research organization, reported on 139 companies of all kinds that were working on new forms of energy. Of them, 28 were petroleum companies. The study found that the petroleum companies tended to concentrate on geothermal energy, oil shale, coal gasification and coal liquefaction. A few oil companies also were active in areas such as hydrogen production, nuclear fusion and breeder reactor research.[20]

Oil companies compete naturally in the development of new energy sources because of their highly developed technology, their skilled energy-trained manpower and their ability to attract investment capital when economic conditions are favorable.

Summary: Actions the Government Can Take

The onshore and offshore lands controlled by the federal government are an essential and substantial resource base for achieving energy security. In order to ensure that those lands will provide the maximum benefits to all Americans, the federal government can take these actions:

- recognize energy production as one of the high-priority uses that can be made of federal lands;
- review and, as necessary, revise federal environmental and land-use rules and policies to ensure that they help maximize domestic energy production;
- permit orderly access to federal lands, both onshore and offshore, for exploration, development and production by responsible, competitive private companies;
- reduce the time required to obtain permits;
- restore the principle of allowing two or more compatible uses of federal lands to co-exist, where that is feasible;
- coordinate the overlapping responsibilities and contradictory regulations of the many agencies involved in the use of the federal lands;
- restore a proper balance between the nation's energy needs and its legitimate concerns for the protection of the environment, wildlife and other values;
- continue to phase out price controls on oil and remove controls over natural gas;
- eliminate artificial governmental restrictions on the distribution of petroleum; and
- adopt sensible tax policies that encourage exploration and production.

* * * * * * *

The Comptroller General of the United States sent a report to Congress on December 7, 1979, about trends in U.S. petroleum and natural gas production. That report, prepared by the General Accounting Office, reached these conclusions:

> Any policy designed to encourage petroleum and natural gas production must have two equally important purposes:
>
> - To provide adequate incentives to drill for new reserves and improve recovery in existing fields.
>
> - To provide incentives to focus new drilling activities in the areas where it is most likely to find new large fields (e.g., the frontier areas of Alaska and the OCS).
>
> To aim for one with little attention given to the other will significantly impair the likelihood of success for any government initiatives to stimulate domestic oil and gas production and thereby restrain oil imports.[21]

* * * * * * *

Positive actions by the government can encourage maximum production of domestic oil and natural gas to help the United States meet a 1990 energy security goal.

Supplement

Examples of What Is Being Done

Further progress in petroleum exploration and production is possible. Here are a number of specific examples of what is being done to increase this country's energy supplies and provide better facilities for processing and transporting oil and gas. Some are substantial, others less so. Cumulatively, these and many other actions are important in reducing the nation's dependence on imported oil:

- The industry's stepped-up activities during 1979 resulted in net increases in proved oil reserves in 10 states. Gas reserves increased in 17 states. California led the list in oil increases with a net gain of 173 million barrels over 1978. North Dakota gained nearly 14 million barrels in oil reserves and Arkansas, more than 13 million barrels. The largest gains in natural gas reserves occurred in Wyoming, Alaska and Kansas.

- In 1979, oil companies paid the federal government a record $1.5 billion in royalties based on offshore oil and gas production. Royalty payments from offshore production have doubled since 1976 and quadrupled since 1972.

- Alaska's Prudhoe Bay field attained its maximum efficient rate of production of 1.5 million barrels a day late in 1979. Companies have invested about $15 billion in that field and in the trans-Alaska oil pipeline over the past decade. Prudhoe Bay now provides 8 percent of the nation's petroleum needs and reduces purchases of foreign oil by $15 billion per year, based on early 1980 prices. The pipeline's capacity was boosted to 1.5 million barrels a day by the installation of additional pumping stations. In addition, a chemical substance is used to reduce friction and allow the oil to flow more smoothly through the 800-mile line. Less than one teaspoon of the chemical per barrel of oil is needed to increase the flow of oil by 150,000 barrels a day. Meanwhile, companies are continuing to explore North Slope areas outside the Prudhoe Bay field for additional supplies of oil and gas.

- Unitized operation, additional drilling and gas injection are helping to increase production from the nation's second largest oil field -- the Yates field in Pecos County, Texas. With more than 1.2 billion barrels in proved reserves, it is second only to the giant Prudhoe Bay field in Alaska. Although several companies own portions of the Yates field, they are operating it as one unit in order to achieve the maximum recovery. During the first eight months of 1978 the field produced an average of 100,000 barrels a day. In 1979 the daily average was 125,000 barrels. Some 60 additional wells were drilled in 1979 to improve the efficiency of the oil production pattern and the distribution of the gas being injected to maintain reservoir pressure.

• Active drilling programs have moved Michigan up from 16th place in oil reserves in 1977 to ninth place in 1979. North Dakota moved up from 14th to tenth place in the same two years.

• In the Gulf of Mexico, a company installed a special $32 million platform in 390 feet of water in October 1979. The platform was built to maintain stability in areas that are prone to mud slides. Beginning in mid-1980, some 22 wells will be drilled to begin producing several million barrels of oil. Meanwhile, in other areas, offshore exploratory drilling is probing waters more than 2,000 feet deep. Engineers are developing concepts for various kinds of platforms that could be used to produce oil or gas from such deep waters.

• A gas injection project getting under way this year is expected to add more than 2 million barrels of oil to future production from the Lake Washington field in Louisiana. The field began production in the 1950s. Over the years natural underground pressures have gradually been used up. The reservoir may stop producing oil within the next two years unless something is done. Gas production from the same field will be injected into the reservoir to build up pressures and force more oil to flow. Engineers estimate this program will extend the life of the oil field 16 years into the future. When oil finally ceases to flow, the gas that has been injected will be produced and sold.

• An intensive program of pressure maintenance and additional drilling has revitalized major oil fields in Wyoming's Big Horn Basin, some of which have been producing oil for 60 years or more. This continuing effort is estimated to have added more than 54 million barrels of net recoverable reserves. A similar program is slowing the natural decline in oil production in southern Alaska's Cook Inlet area.

• A company has resumed heavy-oil recovery operations in Kern County, California, after a lengthy suspension caused by price controls and environmental restrictions. Federal controls on the price of heavy crude oil were removed in August 1979. By February 1980, production had risen to about 65,000 barrels per day. And the company is investing large sums in additional scrubbers and waste disposal facilities to meet environmental requirements.

- A natural gas plant in the Port Hudson field in Louisiana is being expanded. It processed 29 million cubic feet of natural gas daily in 1979. During 1980 the figure is expected to reach 100 million cubic feet a day.

- Companies are making plans to build new pipelines or to expand existing systems to move natural gas to markets from the Overthrust Belt region of the Rocky Mountains. Three preliminary proposals for new pipeline systems have been filed with the Federal Energy Regulatory Commission. They call for moving gas from Wyoming and Utah to California, the Midwest and the Southwest.

- Oil companies are constantly working to develop better refining technology. In March 1980 one company announced a patented high-temperature catalytic conversion process which could result in substantial savings in oil consumption. According to the announcement, the process will make it possible to convert heavy residual oil into gasoline, light distillates and other higher-value specialty products. The company said the new development also will enable refineries to handle a much heavier crude oil than normal or, alternatively, to produce the same yield of gasoline while using 20 percent less crude oil.

- One company reports that from 1975 to 1981, it will have invested $1 billion in a highly efficient complex of refining and transportation facilities in Louisiana. The company has increased the rated capacity of a refinery in that state from 200,000 barrels a day to 255,000 barrels daily, almost tripling the refinery's capacity to manufacture unleaded gasoline.

- Another company has announced plans to upgrade its gasoline manufacturing facilities by building a 40,000-barrel-a-day catalytic reforming unit at its refinery in Port Arthur, Texas. The new unit will increase the company's motor gasoline output by about 475,000 gallons per day and will increase its capacity for producing unleaded gasoline.

- Still another company has begun engineering work on a project to increase a refinery's capability to process high-sulfur crude oils by 95,000 barrels a day. This action will reduce dependence on limited supplies of high-cost, low-sulfur crudes. This is part of the company's $500 million refinery expansion and improvement program, which includes major energy-saving projects at refineries.

- Two oil companies recently announced development of technology for producing a pulverized coal-oil mixture that will remain stable by inhibiting the settling of the coal. This means that the mixture can be stored in tanks for consumption in large boilers.

- One company has obtained permits from officials of Kern County, California, to begin mining diatomaceous earth and testing methods of extracting oil from it. Diatomaceous earth is a fine-grained, porous material formed by the slow accumulation on ocean floors of the shells of minute algae called diatoms, which occur in immense numbers in the sea. Independent consulting engineers have estimated that deposits of this substance near McKittrick, California, may contain 832 million barrels of oil. If commercial processing proves to be feasible, the oil company hopes to recover about 20,000 to 24,000 barrels of petroleum a day by the mid-1980s.

IV. COAL

Introduction

Coal is the most abundant fossil fuel in the United States, accounting for more than three-fourths of known recoverable energy reserves. And the United States has more than one-half of the free world's coal resources.

Coal was first mined commercially in the United States in 1790, near what is now Richmond, Virginia. Within a decade, coal mining also began in western Pennsylvania.

By the mid-1850s, coal had become the country's principal fuel. It remained so until World War II. Now, however, coal provides less than 20 percent of the nation's energy requirements.

For a number of years prior to 1979, the overall demand for coal increased very slowly -- by less than 3 percent a year. However, beginning in early 1979 and continuing into the first half of 1980, coal use has increased substantially. Coal production so far this year is more than 10 percent higher than for the same period in 1979. During the 1980s, coal is capable of providing an even larger share of U.S. energy supplies.

Some people feel that by returning to coal, the nation is taking a step backwards. They argue that coal cannot be mined or burned without harming the environment and that coal is a fuel of limited use. However, the technology already exists -- and is in use -- to mine and burn coal in an environmentally acceptable way. Furthermore, coal can be changed to clean-burning liquid and gaseous fuels that can directly replace petroleum. The nation can expand the use of coal while continuing to improve the quality of the environment.

Abundant and reasonably priced supplies of coal can help boost productivity across the nation's economy. First, if the nation chooses to accelerate the development of its coal reserves, American industries will have greater access to more reliable and less expensive energy supplied by coal. Second, since coal costs less than fuel oil, electricity generated by coal is helping to hold down the cost of electricity to homeowners and other consumers. Finally, money paid for coal stays in the United States and, thus, helps to improve the nation's balance of payments.

Current forecasts from government and industry show coal use in the United States increasing moderately through the coming decade from the 1979 level of 680 million tons a year to around 1.2 billion tons in 1990 -- an increase of about 75 percent. However, if the nation chooses to take advantage of its huge coal reserves and to resolve the political and other problems constraining coal, coal use in this country could double by 1990, to 1.4 billion tons a year or more. Coal would then make a large contribution toward the goal of cutting oil imports by as much as one-half by the end of the decade.

This chapter examines three aspects of coal's potential contribution: resources and reserves, producing more coal and using more coal.

Resources and Reserves

Several different terms are used to categorize the potential and known amounts of coal to be found in the United States. The categories (and their estimated amounts) are:

- Resources -- a staggering 4 trillion tons.
- Proved reserves -- an estimated 1.2 trillion tons.
- Minable reserves -- about 438 billion tons.
- Recoverable reserves -- around 250 billion tons.

The recoverable reserves of coal (defined as coal that can be extracted under current technological and economic conditions) contain the energy equivalent of 1,000 billion barrels of oil. That is more than six times America's total recoverable crude oil resources.

More than half of the minable coal reserves lies west of the Mississippi River -- scattered beneath 20.6 million acres in Wyoming, Montana, Colorado, North Dakota, New Mexico and Utah. The Powder River coal region of northeastern Wyoming and southeastern Montana contains about 60 percent of western minable coal reserves. That region has unusually thick deposits located close to the surface. The seams are up to 120 feet thick and contain an average of about 200,000 tons of coal an acre.

The rest of the nation's minable coal reserves is located in the eastern (Appalachia) and midwestern states. Kentucky, West Virginia, Pennsylvania and Illinois contain 80 percent of these minable coal reserves. About two-thirds of Appalachian and midwestern minable coal reserves are located in relatively thin seams at a depth of 200 feet or more. The average thickness of a coal seam in this area is 5 feet -- a third of the average thickness of a coal seam in the West.

Both eastern and western coal have their own distinct advantages.

Most Appalachian and midwestern coal is bituminous with a high heat content. A ton of bituminous coal contains the energy equivalent of about four barrels of American crude oil. Much of the western coal is sub-bituminous and has a lower heat content by weight. Each ton contains the energy equivalent of roughly three barrels of oil. Thus, even though there are somewhat less minable coal reserves east of the Mississippi, their total heat content is greater than that of western coal reserves.

Western coal, on the other hand, is relatively low in sulfur. The sulfur content of coal in the United States generally ranges from 0.2 to 7.0 percent by weight. Western coal contains on average only 1 percent sulfur by weight and thus emits fewer sulfur compounds per ton of coal burned.

In addition, it is less expensive to produce western coal, which is mostly extracted through surface mining. This is, however, partly offset by its lower heat content and the higher cost of transporting western coal to the major coal users east of the Mississippi River.

Producing More Coal

In the United States, coal production could increase from the 1979 level of about 775 million tons a year to as much as 1.5 billion tons in 1990, according to high supply forecasts reviewed. (Coal production exceeds coal use in the United States because some coal is exported and some is stockpiled for future use.)

Seventy-two percent of U.S. coal production in 1979 was in the Appalachian region and midwestern coal fields. In 1979 Kentucky was the major coal mining state, with nearly 20 percent of total U.S. output -- almost 130 million tons. West Virginia and Pennsylvania were next, together providing 25 percent of coal production.

Although the East and Midwest will continue to provide the bulk of American coal, the western states can make a greater contribution to coal production in the 1980s. In 1979 western coal accounted for 28 percent of total coal production in the United States. Government and industry estimates suggest that western coal reserves could supply around 40 percent of total national production in 1990.

The Powder River coal region is the major coal producing area in the West. The Department of Energy (DOE) expects production in this region to reach about 205 million tons a year by 1985 and nearly 400 million tons in 1990. Thus, the Powder River region could account for at least 30 percent of national coal production by the end of the 1980s.

Mining Methods

Basically, there are two methods of mining coal: underground and surface mining. Underground mining is the major method of producing coal in eastern states -- particularly West Virginia and Virginia. This method is used when coal is found at depths of 150 feet or more below the surface.

The two techniques for mining underground coal are "room and pillar" and "longwall." The room and pillar technique is by far the more prevalent, currently accounting for 80 percent of underground mining output. This technique consists of cutting panels ("rooms") into the coal seam. As the coal is extracted, pillars of coal are left in place to hold up the roof. Initially, as much as half of the coal remains underground. Later on, some of the pillars are removed, increasing the recovery rate.

The longwall technique uses hydraulic-powered supports to prop up the roof of the mine while large mechanically driven shearers cut away at the coal. Since there are no pillars, this technique extracts a greater amount of the coal than the room and pillar technique. After the coal is recovered in longwall mining, the supports are removed and the roof is allowed to collapse safely. This technique is used where the geological formation above the coal seam allows controlled and uniform settling.

The second method -- surface mining -- is used where coal is found up to 150 feet beneath the surface. In such instances, it is safer, more efficient and more economical to employ the surface-mining method. Currently, surface mining accounts for about 60 percent of coal production nationwide. Surface mining has become the major method for extracting coal in the West and in some eastern states.

In surface mining, the overburden (soil and rock) above the coal seam is removed so that shovels and scrapers can extract the coal. After the coal is mined, reclamation of the land begins. The pit is refilled with the overburden material and soil is spread on top of the overburden so that the land is returned to its approximate original contour as required by the Surface Mining Act of 1977. The area is then reseeded or replanted with appropriate forms of vegetation. (It should be noted that the requirement that reclaimed lands be returned to their approximate original contour has proved to be burdensome especially for contour operations on steep slopes. The requirement has particularly impacted on steep slope operations in various Appalachian states.)

The Role of Public Coal Reserves in Future Production

Most eastern coal reserves are privately owned. On the other hand, the federal government owns about 60 percent of western coal reserves. An additional 20 percent of privately owned western coal is dependent on access across federal lands for its production.

Federally owned coal is concentrated in six western states: Colorado, Montana, New Mexico, North Dakota, Utah and Wyoming. These six states produced 71 percent of all western coal in 1977. Federally owned coal accounted for less than half (43.7 percent) of total coal production in these states and for only 8 percent of national coal output.

Because of its ownership or control of about 80 percent of western coal reserves, the federal government has almost absolute control over which reserves are and are not produced west of the Mississippi, even on some private lands. The government has, in fact, exercised that control.

In 1971, the Department of the Interior (DOI) imposed a moratorium on coal lease sales. At that time, it undertook to develop a new program that involved more pre-lease analysis. It also ordered that an environmental impact statement on the leasing program be developed. The new program, adopted in early 1976, was challenged in court. The court, in 1977, directed that DOI prepare a revised environmental impact statement. Instead, DOI elected to create an entirely new program. This program was adopted on June 1, 1979. Three days later, the DOI secretary announced specific areas and targets for the first lands to be leased, beginning in January 1981 -- nearly 10 years after the moratorium was imposed.

Other government agencies (DOE, the Department of Justice, the Council on Wage and Price Stability) -- as well as coal industry officials -- have raised a number of analytical and factual questions about the coal leasing targets. The Council on Wage and Price Stability, for example, contends that DOI has overestimated the amount of coal that could be produced from existing leases and has failed adequately to take into account the impact of its own new land use planning regulations on new and existing leases.[1]

In a speech earlier this year, National Coal Association (NCA) president Carl E. Bagge expressed the coal industry concerns over the new DOI program. He said:

> In setting the targets for new leasing, the Department assumed production from existing operations would continue at existing levels. It is possible, however, that as many as half of the tracts now under consideration for new leasing would in fact be needed to sustain existing production. What this means is that in setting the new leasing target numbers, the same coal may have been counted twice. Even if all of the coal now under consideration in this region were able to be leased, it is likely that Interior's targets would not be met.[2]

The NCA president pointed out that, under the criteria developed by DOI, any interested citizen may file a petition to designate lands unsuitable for mining. He added: "Based on little more than a slip of paper, interested parties can stop mining immediately."[3]

Bagge summed up the industry's concerns over DOI's plan by stating that it "will limit supply and increase coal prices, thereby reducing or eliminating economic incentives to use coal in preference to oil or natural gas."[4]

Private Land Ownership and Coal

In the western states, private property owners can often prevent access to federal coal reserves. These private lands are interspersed with federal lands in a complex checkerboard pattern. As a result, private property owners can complicate leasing of federal coal beneath their land and could prevent mining of federal coal on nearby lands. Through public hearings, they can argue that coal mining would have an adverse impact on the community and the environment. While not all private land-owners are against development, the veto power of a few in a checkerboard area is sufficient to block access to coal resources.

However, some federal resources do not face this problem. Government owns both the surface and subsurface rights for some western coal lands that have not yet been leased. In other instances, coal mining companies own the surface rights to considerable acreage in coal areas and are waiting for the government to lease the reserves beneath these lands.

Coal Mining and the Environment

Legitimate environmental concerns can be resolved so that production of coal can be increased.

Past coal-mining practices -- continued under laws and regulations that existed at the time -- did have an impact on the environment in many instances. Today's laws, however, do not permit repetition of these past practices. In addition, recent experience demonstrates that coal is being extracted in an environmentally acceptable way. Through programs of land reclamation, pollution control and water conservation, government and industry can continue to protect the environment while greatly expanding coal production.

Land Disturbance and Reclamation. Local land owners, outdoor recreationists and others are especially concerned about the possible impact of surface mining. However, strict federal and state laws now require reclamation of mining areas immediately after production.

In areas where coal seams are unusually thick, productive mining can take place on a relatively small amount of land. In the Powder River area, for example, miners can extract approximately 200,000 tons of coal on average while disturbing only about one acre of surface area.

A good example of land reclamation is the McKinley coal mine near Gallup, New Mexico. The mine is operated by the Pittsburg and Midway Coal Mining Company, a subsidiary of Gulf Oil Corporation. Reclamation begins almost as soon as coal is removed from a particular tract. Bulldozers regrade the land to the approximate original contour, and the topsoil is prepared with fertilizers, seeded and mulched.

In many instances, reclamation has improved the quality of the land after it has been surface-mined. At the McKinley mine, for example, New Mexico State University scientists have developed seed mixtures that increase the productivity of the forage for sheep grazing in the area. Furthermore, because the McKinley area is dry, with only 12 inches of rain a year, water control and conservation measures are incorporated into reclamation.

Water Availability. In the semi-arid West, concern about water supply could be a major constraint on commercial coal development. Some concern has also been expressed about the impact of coal mining on aquifers (water-bearing rock formations beneath the ground). Studies indicate, however, that the water supply in the West is sufficient for mining and other activities.

This fact is borne out in a January 1980 report of the General Accounting Office's Community and Economic Development Division. The report, "Water Supply Should Not be an Obstacle to Meeting Energy Development Growth," concluded that "adequate water is available for energy development through at least the year 2000." The report called "unfounded or outdated" predictions that energy development in the West would virtually exhaust the water supply there.[5]

A case in point are the activities in the Powder River coal region, which draws water from the Yellowstone River. Roughly 11 percent of the river's flow is used for agricultural irrigation. By contrast, all mining activity in the region uses less than 0.04 percent of the river's water supply. The Water Resources Council estimates that if coal production in the region increased to the unusually high level of 232 million tons a year by 1985, mining would use 1 percent of the river's average annual flow.[6]

In brief, there is adequate water to support mining and other activities in the West. Careful operation of mining equipment, as well as careful allocation and conservation, will ensure that all users have enough water.

Air Pollution. Coal mining without proper controls could -- like other major human activities -- raise clouds of dust that do not resettle quickly in arid windswept areas of the West. These tiny particles of coal and soil if left uncontrolled can affect visibility and can cause difficult breathing for people near surface mines.

However, mining companies have taken steps to control dust pollution by spraying roads and exposed areas with a chemical binder (surfactant) that holds the surface together. They also attempt to reclaim mined areas immediately after the coal is removed in order to prevent erosion and to reduce the possibility of pollution from dirt and other particulate matter.

Using More Coal

Substantial increased use of coal can be achieved through resolution of a number of regulatory, economic and environmental questions. The principal opportunity for using more coal in the 1980s and beyond is in the electric utility industry.

Utility and Industrial Use of Coal

Electric utilities are, and will continue to be through the 1980s, the major consumers of coal in the United States. They now account for over 70 percent of the demand for coal. For all of 1979 coal-fired plants consumed about 530 million tons of coal and generated approximately one-half of the nation's electricity. By 1990 coal consumption by utilities could increase to around 1 billion tons a year.

Nearly all new power plants planned by utilities will either be nuclear or coal-fired. No new base-load oil or natural gas plants have been ordered by utilities since 1974.

Electric utilities' plans call for bringing into service 33 new coal-fired units in 1980 and a total of about 275 new units over the next 10 years. By 1990 these new units would require an estimated 400 million tons of coal a year if the plants are not delayed. This amount of coal is equivalent to more than 4 million barrels a day of oil.

The National Coal Association estimates:

> Each 100 million tons of additional coal that is used in lieu of foreign oil will permit the United States to avoid importing 400 million barrels of oil and avoid the outflow of over $12 billion.[7]

In addition to avoiding increased oil imports, coal use can help reduce imports in several ways. First, existing coal-fired plants can be operated at higher capacity. The electric utility industry has already moved in this direction mainly as a result of sharp increases in world oil prices in early 1979. Utilities with extra coal-fired capacity have sold power to systems with oil-fired capacity. This practice -- called "wheeling" -- accounts in part for the increased use of coal by utilities and the reduction in their use of oil over the past year or so. Similarly, the new coal-fired plants being built by utilities can be used to reduce reliance on existing oil and gas-fired plants.

Second, some existing power plants and large industrial boilers now using oil or natural gas can be converted to coal.

Finally, the use of coal-oil mixtures should allow increased coal use and reduced oil use in plants that cannot be converted to coal alone. The technology and economics of this step have recently improved, especially since the 1979 increase in world oil prices. Demonstrations and tests that had been under way in Florida and Massachusetts are now moving into commercialization.

The potential oil-import savings from steps such as these are sizable. In 1979 utilities burned 1.3 million barrels a day (mbd) of residual oil, 0.1 mbd of distillate and 1.7 million barrels a day of oil equivalent (mbde) of natural gas. (Residual oil is heavy fuel oil, used mainly in utility and industrial furnaces and boilers. Distillate oil consists primarily of home heating oil and diesel fuel.) In addition to utilities, large industrial and commercial boilers -- including some 12,000 boilers representing 42 percent of total boiler capacity -- used about 0.3 mbd of residual oil, 0.1 mbd of distillate and 1.2 mbde of natural gas.

So, total use in utilities and large industrial and commercial boilers in 1979 was 1.6 mbde of residual oil, 0.2 mbd of distillate and nearly 3 mbde of natural gas. Of these amounts, coal could potentially replace the 1 mbd of residual oil we now import and perhaps 1 mbde to 1.5 mbde of natural gas that can be used in other sectors to replace oil.

Thus, in an accelerated case for 1990, the total potential oil import reduction from increased coal use could be as high as 2 mbd to 2.5 mbd.

The actual level achieved by 1990 will depend on several factors, particularly the speed with which new coal-fired plants can be brought into service. Another factor that has received much attention recently is conversion of existing plants to coal. The units identified by the Department of Energy for conversion to coal could use about 20 million to 30 million tons a year of coal, or the equivalent of as much as 0.3 million barrels a day of oil. The National Coal Association's upper estimate for potential coal conversion is 20 million tons a year.[8]

Legislation. Congressional actions to convert electric power plants from oil or gas to coal resulted in the passage of two pieces of legislation in recent years. A third coal-using measure has been under consideration by Congress.

In 1974 the Energy Supply and Environmental Coordination Act (ESECA) became law. It sought to prohibit certain power plants that can burn coal from using petroleum and to require that new fossil-fuel fired boilers be designed to burn coal. The goal of this law was to replace 0.3 mbde of oil and gas, although its effect has been small.

In 1978 the Powerplant and Industrial Fuel Use Act (FUA) extended the provisions of ESECA and set a short-term goal of replacing 0.4 mbde of oil and gas by 1985. FUA seeks to reduce the use of natural gas and oil in new electric power plants or industrial boilers and, after 1989, the use of natural gas in existing power plants. FUA also requires that all new power plants be capable of burning coal or alternative fuels. However, the act permits many exemptions and exceptions to its provisions.

In 1980 President Carter introduced the Powerplant Fuel Conservation Act. He indicated that the goal is to displace 0.75 mbd of oil and 0.25 mbde of natural gas by 1990. This legislation covers conversion of existing plants to coal, as well as the other steps by which coal will displace oil and gas.

Some utilities are proceeding with conversion of existing facilities to coal, mainly because of higher oil prices on the world market and the greater security of supply when using coal. These conversions depend upon certain environmental regulatory exemptions authorized by FUA.

Achieving higher levels of coal substitution through conversion of existing plants and use of new coal-fired plants will depend on government attention to environnmental regulatory conflicts and financial obstacles.

Regulatory Conflicts. Environmental regulations presently create obstacles to coal conversion and other oil and gas displacement measures.

In a recent report, the Department of Energy noted, with respect to existing power plants:

> Environmental considerations will generally not prevent conversion of units, but compliance with environmental laws and regulations can delay the conversion process or raise conversion costs so high by requiring the upgrading or installation of pollution control equipment that conversion becomes economically unattractive.[9]

Many of the legislative and regulatory restrictions have been based on assumptions or unsubstantiated conclusions that have later proved to be erroneous.

Both government and industry recognize that, before the United States can make much greater use of coal, the regulatory process will have to be streamlined and made more flexible. If Americans choose this route to regulatory reform, they can benefit further from the burning of even greater amounts of coal. And more coal can be burned without damaging the environment or imposing economic hardships on utilities and consumers.

Financial Obstacles. Financial problems have discouraged some utilities from converting existing units from fuel oil to cheaper coal. Conversion may require a multi-million dollar investment for new coal-handling equipment, pollution control devices and changes in boilers. Many utilities -- with their regulated rates of return -- have cash flow or capital availability problems that create difficulties in acquiring loans for improvements. The cost of conversion imposes a debt that many utilities have difficulty in handling.

The President's Commission on Coal, in its March 1980 report, pointed specifically to this problem:

> State regulatory utility commissions can hinder utility coal conversion and oil replacement efforts by not granting adequate rates to utilities to cover the front-end capital costs associated with switching to coal. These construction costs may raise electricity rates in the near term, but over its lifetime the coal-converted unit will cost consumers less because of the fuel savings from coal.[10]

The result is that some utilities cannot switch to coal as quickly as they would like to. The president and the utility industry, in particular, have cited the economic and regulatory obstacles to even greater coal use as justification for the Power Plant Fuel Conservation Act. They have suggested that government provide direct grants and other economic incentives to utilities to speed the conversion of existing coal-capable plants from oil to coal.

Transportation

Before the country can benefit from a further increased coal supply, the nation will need to more fully develop efficient, economical and environmentally acceptable systems for moving coal.

Some of these systems, such as railroads and waterways, already exist but need to be further improved or expanded. Others -- particularly coal slurry pipelines -- need to be constructed.

Upgrading the Railroads. Trains are the major coal carriers in the United States. The National Coal Association says that trains carry 65 percent of coal supply and that most of this coal has no alternative economic means of transport. The coal association expects that trains will haul 500 million tons of coal in 1980 -- 70 percent of coal transported -- and continue to carry the bulk of coal through the next 10 years.[11] In order to handle that high volume of coal safely and efficiently, however, the railroads will need to upgrade some parts of the system.

A second potential problem is the availability of sufficient coal cars and locomotives. Fifty percent of the coal carried by rail moves on unit trains. Each train can carry 10,000 to 12,000 tons of coal -- 100 tons in each "hopper" car. (A hopper car is an open car with trap doors for unloading coal when it reaches its destination.) In 1976 one railroad company used 55 unit trains; by 1985 it will require as many as 200 units.

Some concern has been expressed that the capital required to expand the capacity of the railroads may not be at hand. The authors of Energy Future, the report of the energy project at the Harvard Business School, have pointed out that "the new demand on railroads to carry coal places enormous stress on their capital expenditure requirements for hopper cars, locomotives, physical plant improvements, and maintenance facilities."[12]

However, both government and industry groups believe that, if demand for coal expands at the projected rate, the capital will become available for upgrading the railroads and other parts of the transport system. In the May 1980 report of the World Coal Study (WOCOL), government and industry participants concluded that "the amount of capital required to expand production, transport and user facilities to triple the use of coal [by the year 2000] is within the capacity of domestic and international capital markets."[13]

Water and Truck Transportation. The WOCOL study also pointed out the need for alternatives to railroads for inland transport. In the United States, barges and other vessels currently carry 10 percent of all coal supply; trucks transport roughly 13 percent. A small amount is moved by pipeline and the rest of the coal is consumed at the mine for generating electricity.[14]

Certain constraints, however, limit the expansion of waterways and trucking for coal transport. Although waterways (where they exist) provide the most efficient and least expensive means of moving coal, there are no waterways serving many areas where coal needs to be moved. The high cost of sending coal by truck discourages the use of motor vehicles to haul coal over long distances.

Coal Slurry Pipelines. The major alternative to the railroads for long distance transport is the coal slurry pipeline. In a slurry pipeline, the coal is first crushed to a small size and then mixed with water. The coal-water mixture (or slurry) is then sent through underground pipelines.

Only one coal slurry pipeline is currently in commercial operation in the United States, transporting nearly 5 million tons of coal annually 273 miles from northeastern Arizona to southern Nevada. However, a considerable number of other pipelines have been proposed for construction in various sections of the country. The longest line, which would pipe coal 1000 miles from Wyoming to Louisiana could be in operation by 1985. (Consolidation Coal Company has installed a slurry pipeline system at its Loveridge, West Virginia, mine to transport coal from underground to the surface and then overland to the preparation plants.)

Government and private studies indicate that in certain situations, coal slurry pipelines are the most economical and the most environmentally acceptable way to transport coal. However, the issues of rights of way across other transport routes and acquisition of water rights are delaying construction of slurry pipelines.

Companies need to acquire rights of way in order to construct coal slurry pipelines. Construction of a pipeline invariably requires the crossing of routes on lands controlled by other transport systems -- principally the railroads. Many states have already enacted eminent domain laws which provide access to such lands; other states do not. Without eminent domain, negotiations for pipeline rights of way could lead to disputes that can be resolved only through time-consuming and costly litigation.

Proposed eminent domain legislation has been before Congress for several years. The legislation would extend to the federal government the power of eminent domain over slurry pipeline routes. This would put slurry pipelines on a par with pipelines carrying other commodities which already benefit from eminent domain laws. Both DOE and DOI have testified in favor of this legislation and have stated that coal slurry lines are efficient and environmentally sound.

The question of water rights is an especial concern in western states. Local authorities may be reluctant to grant water rights to pipeline companies when the authorities must also meet the demands of farmers, ranchers and small industries. Roughly a ton of water or other fluid is needed for every ton of coal. While that sounds like a great deal of water, slurry pipelines use considerably less water than generating plants or coal gasification plants located at mines.

The General Accounting Office's January 1980 report on water availability indicated:

> Since coal slurry lines can use water that is too contaminated or too expensive for other purposes, the technology should not have much impact on other water consumers.[15]

Thus, while transportation does pose some obstacles to coal use, these obstacles can be overcome so that more coal can be moved from the mine to the marketplace.

Coal Use and the Environment

Another challenge to increased coal use is in using the nation's abundant coal supplies in an environmentally sound way based on practical environmental laws and regulations.

Burning coal in the old way -- without modern pollution controls -- would cause significant air pollution and could have some adverse impact on living things on land and in water. Today, pollution control devices -- now in place on most coal-burning equipment -- have sharply reduced the emission of pollutants into the atmosphere. However, total control of emissions from coal-fired boilers has not yet been achieved. Pollutants from coal burning include sulfur dioxide, nitrogen oxides, particulates and carbon dioxide. (Particulates include dust, soot, fly ash and other tiny particles of solid and liquid matter.)

Current Technology. The technology to limit emissions of these pollutants exists and is being used by many facilities that burn coal. This technology includes the following techniques and equipment:

- Coal washing. Coal can be cleaned before burning to remove some of the impurities contained in the coal.

- Electrostatic precipitators. This equipment reduces particulate emissions. The newest precipitators are considered to be 90 percent to 99 percent efficient in removing the larger particles.

- Stack gas scrubbers. These devices control the emission of sulfur oxides. As of the end of 1979, 62 scrubbers were in commercial operation, with an additional 39 scrubbers under construction. According to the Environmental Protection Agency, scrubbers installed in large electricity-generating power plants remove up to 95 percent of sulfur emissions.

These pollution control techniques are being made more effective. And further environmental gains will occur as more coal-powered facilities are able economically to install these devices.

Other Environmental Concerns. Public attention has recently been given to several broader environmental concerns that could impact on the greater use of coal. These involve "acid rain" or "acid precipitation," the "greenhouse effect" and solid waste disposal.

a. Acid Rain. Some scientists contend that rainfall in some parts of the world (including areas of the United States) has become more acidic in recent years. Other scientists question these findings. They contend that available data on this phenomenom are not adequate to lead to the conclusion that acid rain is a serious problem. Disagreement also exists over the causes and effects of acid precipitation.

According to Douglas Costle, EPA administrator, "the major component of eastern acid rain is sulfuric acid (due in large part to pollution from sulfur emitting stationary sources, primarily coal-fired power plants)." He added that advanced coal washing techniques could reduce the severity of the problem. Costle claims that, when acidic rainwater falls to earth, it destroys fresh water ecosystems, damages soil, erodes stone buildings and monuments and causes poor visibility. He adds that acid rain is also suspected of damaging important crops, reducing forest growth and contaminating drinking water.[16]

This view is not shared by industry specialists. The Electric Power Research Institute (EPRI), in a recent publication, stated:

> Like a classical mystery story, the train of events involving acid rain [is] complexly interwoven and the clues that appear are extremely subtle. Existing evidence is often as erratic as an April shower.[17]

The publication added:

> Whether or not the rain is actually becoming more acidic remains unclear ... long-term trends of acidity can only be determined from data collected for several years at identical sites. Yet, for the period from the mid-1950s to the early 1970s, there were only two stations from which such observations could be obtained. At one site the acidity of rain increased; at the other, it decreased.[18]

Industry specialists also say that the relationship between acid rain and coal-burning utilities has not been proven. Dr. Ralph Perhac, director of EPRI'S environment assessment department, states there is still no clear idea of how utility emissions relate to acid rain. He adds that, if the utilities' role is not clearly defined, then "we don't know how changing [the emissions] might help.[19]

William N. Poundstone, executive vice president of Consolidation Coal Company, has stated: "There is no question that rain is acidic. But the relative degree of acidity arising from natural and from man-made sources is largely unknown." He has pointed out that there are many natural as well as man-made sources of the compounds (carbon dioxide, sulfates and nitrates) that contribute to the acidity of rain. The natural sources include lightning, volcanos, sea spray and the organic decay of vegetation.[20] Citing a National Research Council report, Poundstone added:

> The percentage of total atmospheric sulfur compounds from <u>natural</u> sources -- worldwide -- is believed to be more than 70 percent, and the percentage of total atmospheric nitrogen compounds from natural sources is believed to be in excess of 90 percent.[21]

In commenting on the claim that coal-burning facilities are a major cause of acid rain, Poundstone cited a report in <u>Scientific American</u> (October 1979). He remarked:

> An examination of the amount of coal burned in this country during the same time that acid rainfall allegedly has increased -- 1955 to 1980 -- reveals that little if any more total sulfur dioxide is today

> being emitted into the atmosphere as a result of coal burning.[22]

Elsewhere, Poundstone has noted:

> If the control of sulfur dioxide emissions, especially from coal-burning power plants, were truly the answer to this problem, then the rain should be becoming less acidic instead of more acidic because the nationwide levels of sulfur dioxide have been going down during the last 10 years.[23]

While the various specialists do not agree on the causes and extent of the acid rain problem, they do agree that not enough is known about acid rain and that further scientific research is needed.

In August 1979, President Carter announced a comprehensive interagency research program funded by $10 million in the first year to determine if acid rain is a problem and, if so, how serious a problem. Subsequently, a measure calling for a study on the causes and effects of acid precipitation throughout the United States was made a part of legislation creating a synthetic fuels corporation. That legislation was passed in June 1980.

In late 1979, EPA stated at a House committee hearing that three to five years of research would be required before a judgment could be made regarding whether new air quality regulations would be required to control acid rain. Recently, EPA and DOE joined together to sponsor an acid rain research program.

Other acid rain research projects are under way or in the planning stages by industry and independent research organizations.

EPRI, which is heavily involved in acid-rain research, indicated:

> Until a generally accepted understanding of acid rain is reached, however, new regulations might prove self-defeating. Preliminary results from EPRI studies show, for example, that even drastic cuts in power plant emissions might have little noticeable effect on the incidence of acid rain.[24]

Poundstone summed up the importance of acid rain research when he said:

> Reasonableness demands that we take the necessary time -- not a day more or a day less -- and spend the necessary money to find the answers before we inflict the additional staggering costs of a control strategy on a public whose welfare is being threatened by higher prices on every front.[25]

b. Greenhouse Effect. When coal (or any other fossil fuel) is burned, carbon dioxide emissions occur. In itself, carbon dioxide is harmless -- nature itself is a major source. However, some scientists believe that large concentrations of carbon dioxide in the atmosphere can, in time, cause climatic changes -- specifically, higher temperatures worldwide (the "greenhouse" effect).

Other scientists are more sanguine about the presence of carbon dioxide in the atmosphere. Some scientists, including Dr. Carl Sagan, Cornell University astronomer, see a cooling phenomenon as counteracting the greenhouse effect. He says that such activities as forest clearing, livestock overgrazing and increased land aridity have, over the centuries, resulted in a higher degree of reflected sunlight from the Earth's surface. The more sunlight the Earth reflects, the cooler it gets.[26]

Members of the World Coal Study concluded that present knowledge of carbon dioxide effects on climate "does not justify delaying the expansion of coal use." The WOCOL report added, "It may happen that some effects of CO_2 [carbon dioxide] will become detectable on a regional and global scale before the end of the century." This finding, WOCOL pointed out, is consistent with the authoritative statement on the carbon dioxide question issued by the World Climate Conference in 1979.[27]

c. Solid Waste. A third problem is the disposal of the solid wastes that result from controls on sulfur emissions. Most scrubbers installed in the past by large coal-burning facilities to reduce emissions of sulfur compounds produce a solid waste called "sludge." The sludge coming out of the scrubbers has to be disposed of in an environmentally acceptable way. This is proving difficult because of a lack of sufficient disposal sites.

However, scrubbers are now available that do not create sludge but instead produce usable and marketable sulfur solutions and pure sulfur.

The Environmental Problems Are Solvable. Government and industry leaders alike agree that coal can make a much greater contribution to the nation's future energy needs in an environmentally acceptable way. Even environmental officials in government say that environmental problems associated with the burning of coal in large quantities are not insurmountable.

EPA administrator Douglas Costle said in an April 1980 Congressional hearing that "conversion [to coal] can be done in a way that will prevent environmental degradation."[28]

At an earlier hearing, EPA deputy administrator Barbara Blum stated, "We believe that coal can be burned cleanly and that it must be burned widely in our country." She added,"We at EPA believe that we can burn as much coal as is necessary -- in a clean way."[29]

The President's Commission on Coal, after a two-year study, concluded that, "the problems associated with coal use can be overcome, and that this nation must begin to rely more heavily on its vast coal deposits to reduce much of our intolerable dependence on imported oil." The commission's report said:

> The Commission's careful conclusion is that a program of replacing oil with coal in compliance with the Clean Air Act will not increase emissions of environmental and health-related pollutants and will cause, at most, minimal increases in atmospheric concentrations of carbon dioxide.[30]

The 1980 report of the World Coal Study concluded: "Coal can be mined, moved, and used in most areas in ways that conform to high standards of health, safety and environmental protection."[31]

DOE takes the position that we can increase the use of coal and still meet present environmental standards with no significant damage to the environment.

Finally, the nation's coal producers have said that "coal can be mined safely, delivered cheaply and burned cleanly." In a letter to the president in March 1980, they urged that the government take actions "to reduce the adverse impact of existing federal and state air quality requirements and permit increased use of coal in industrial and utility facilities. These steps could be taken without violating air quality standards set to protect public health."[32]

New Approaches to Environmentally Acceptable Coal Use. Meanwhile, many different new processes and techniques are being demonstrated in pilot plants or are undergoing testing in research facilities. These processes and techniques can be grouped into four categories:

a. Coal cleaning processes -- which remove most of the impurities from coal before it is burned;

b. Emission-control devices -- more sophisticated equipment which removes pollutants during the process of coal combustion;

c. New coal technologies -- which, when developed, will use coal efficiently and without environmental hazards; and

d. Synthetic fuels from coal -- the conversion of coal to a liquid or gaseous form. The conversion process removes the potential pollutants and produces a fuel which can be used in the combustion equipment already in place.

Various processes are being tested and demonstrated in all four categories. These experiments are being conducted by universities, research institutes, private companies and the federal government.[33]

a. Coal-cleaning processes. Washing coal, before it is shipped to utilities and industrial plants, is already in use in some areas. In its simplest terms, the process consists of separating the coal from its impurities through washing. However, according to EPRI, "existing technologies, some admittedly primitive, must be improved upon, and other, newer technologies developed."[34] These would include methods to effectively dispose of the wastes resulting from coal cleaning.

As with other experimental and emerging technologies, coal-cleaning offers a double benefit. Removing impurities (ash-producing minerals and sulfur) and drying surface moisture results in a coal which emits fewer pollutants and which has a higher heat content.

EPRI estimates that, by 1985, over one-third of the coal used by the utility industry will be physically cleaned. "Cleaning technologies," EPRI says, "will permit using the large coal reserves of the United States interior and Appalachian region that otherwise could not be used under sulfur oxide emission requirements."[35]

b. More sophisticated emissions-control devices. Experiments are under way to make existing pollution-control devices even more efficient. For example, EPRI and EPA are studying ways to improve the performance of electrostatic precipitators to control emissions of very fine particles emitted during coal burning.

A device that has been used to control emissions from other types of facilities is also being adapted to reduce or remove emissions of fine coal particles. The device, called a "baghouse," consists of thousands of bags (one foot in diameter and 30 feet long) generally made of glass fiber. The baghouse traps the particles, preventing them from being emitted into the atmosphere. Use of baghouses on a small scale has produced efficiency figures of 99.8 percent removal of total particulate mass and better than 99 percent for the smaller particles, according to EPRI. Baghouses are not in general use, however, because of the "absence of thorough cost and reliability data," EPRI states.[36]

Further improvements in current technology can be expected as coal use increases.

c. New coal technologies. A number of experimental processes are being tested to burn coal directly in a more environmentally sound way.

One process is called "solvent refined coal" (SRC). This refining process results in an environmentally acceptable fuel from coal originally high in sulfur and fly ash. In the SRC process, coal is mixed with a solvent and treated with hydrogen at high temperature and pressure. Several oil companies are involved in SRC research and development. For example, Gulf Oil Corporation, through its subsidiary mining company, has operated an SRC pilot plant at Ft. Lewis, Washington, since 1974.

A second process, called "fluidized bed" combustion, removes the sulfur from coal as it is being burned. The fluidized bed system uses a stream of air to partially suspend a bed of limestone particles. Coal that has been pulverized is then injected into the bed. According to engineers at the MIT Energy Laboratory, fluidized bed combustors combine high combustion efficiency with levels of sulfur and nitrogen oxide emissions low enough to meet EPA standards.[37] The first coal-fired industrial fluidized bed generator to go into commercial operation in the United States was installed at Georgetown University, in Washington, D.C. The unit was started up in July 1979 and now supplies the university with all of its steam requirements.[38] Several electric utilities have also experimented with fluidized bed combustion in some boilers.

The MIT Energy Laboratory is experimenting with a modified fluidized bed combustion process as a means of achieving further reductions in nitrogen oxides emissions. The modified system is undergoing tests in the laboratory's pilot-scale experimental combustor.[39]

Demonstrations and tests are also under way to further develop the technology to produce and use coal-oil mixtures, as noted earlier. The widespread use of this approach commercially, however, will require government approvals for use of coal-oil mixtures and for siting and building the facilities to produce and transport the mixtures.

d. Synthetic fuels from coal. The basic technology for changing coal to liquid or gaseous form has been available for some 50 years. One country -- South Africa -- already has an extensive capability to liquefy coal. One plant has been in operation in South Africa for several years; a second, much larger plant, went into operation early in 1980. When this second plant is fully operational, oil from coal will account for about 35 percent to 50 percent of that country's total petroleum consumption.

Price controls on domestic oil and natural gas and the low price and ready availability of imported oil, until recently, discouraged the development of synthetic fuels from coal in this country. But this situation is changing.

Specific examples of synthetic fuel research and development by petroleum and other companies are cited in Chapter VI, "Synthetic Fuels and Renewable Energy."

It is generally agreed that greater use of coal will spur the development of the technology needed to burn coal more efficiently and in a more environmentally sound way.

Conclusion

The Chairman of the National Coal Association, R. E. Samples, recently summed up the increasingly important role that coal can play in solving the nation's energy problems. He said:

> The nation has abundant coal reserves. We have the productive capacity in place to provide additional coal. We have the ability to expand capacity when demand warrants.
>
> We have an abundant labor supply. Labor and management in the coal industry are working together as

never before to solve common problems. There are greatly improved chances for labor peace in the years ahead.

We have vigorous competition in the coal industry which has and will continue to contribute to the stability of coal prices.

We have the ability to mine and use coal in a safe and environmentally acceptable manner.

Opportunities are available now to increase the use of coal and reduce dependence on imported oil.

The problem ... is the web of government policies and requirements that unnecessarily increase the cost of producing, transporting and using coal, and that holds demand below coal's full potential.[40]

National Coal Association president Carl E. Bagge sought to put the problem of government policies and requirements into perspective. In May 1980 he stated:

We want to emphasize that action to reduce obstacles to an increased contribution from coal does not require abandoning environmental, health and safety or other goals or wholesale overhaul of laws and regulations. Instead, the need is to identify and adjust those requirements which are more stringent than necessary, which introduce unnecessary delays or costs, or which do not reflect the best balance among national objectives.[41]

Finally, the World Coal Study report states:

Building a bridge to the energy sources and supply systems of the next century -- whatever they turn out to be -- is of crucial importance. We believe that coal can be such a bridge and that it will also continue to serve a vital role into the longer-term future.[42]

The United States has nearly 32 percent of the estimated worldwide coal resources and more than one-half of the free world coal resources. We thus have a unique opportunity to help reduce our continued high dependence on oil from insecure sources and to reduce political and economic pressure on world oil markets.

This opportunity presents an enormous challenge to work toward solutions to the environmental, political and economic obstacles that presently hinder the increased production and use of coal in the United States. But this is a challenge that can be met.

V. NUCLEAR ENERGY

The importance of nuclear energy has often been cited by President Carter. In a comment on the report of the Presidential Commission on the Accident at Three Mile Island, he said:

> We cannot shut the door on nuclear energy. Every source of energy, including nuclear power, is critical if we are to free our country from its overdependence on unstable sources of high-priced foreign oil. We do not have the luxury of abandoning nuclear power or imposing a lengthy moratorium on its further use.[1]

What is the outlook for nuclear power in 1980, one year after Three Mile Island?

There are currently 72 reactors with operating licenses. Another 89 reactors have construction permits and two have limited work authorizations. Another 19 are on order.

Nuclear plants provided about 12 percent of the electricity produced in the United States last year, despite shutdowns of several reactors mandated by the Nuclear Regulatory Commission (NRC). The output of the plants was about 1.3 million barrels a day of oil equivalent (mbde).

Most analysts expect that substantial gains will be made in the next decade. If all 110 plants under construction and ordered are completed during the decade, the nuclear contribution to the nation's energy would increase to nearly 4.5 mbde. More likely, some of these plants will be delayed until after 1990. Even so, the experts expect that nuclear energy can still about triple -- to about 4 mbde by the end of the decade.

That is a realistic achievement that can be met if currently completed nuclear plants are in operation soon and if plants with construction permits are completed as scheduled. And it can be achieved within the framework of new safeguards developed after the Three Mile Island accident.

Thus, nuclear energy can have a key role in meeting the goal of reducing imports by as much as 50 percent by 1990.

Background

Nuclear energy has a substantial role to play in the overall energy picture because the United States has been shifting toward reliance upon electrical power. Only about 12 percent of the nation's electric power is now being generated with nuclear fuel, so there is considerable room for growth.

In the early 1970s, forecasts for nuclear energy were quite optimistic, predicting levels of 7 mbde or higher for the late 1980s. However, doubts about nuclear's future soon began to surface, and the forecasts began dropping. After Three Mile Island, the future appears even less certain. Opponents of nuclear energy are calling for the shutdown of all existing nuclear power plants. Later this year, citizens of the state of Maine will vote on a proposal to prohibit the generation of electricity by nuclear power. If passed, that measure would shut down the Maine Yankee plant near Wiscasset. Whatever the outcome, the initiative probably heralds a number of challenges in other states in the years ahead.

Even with such problems, current forecasts suggest that nuclear power will continue its growth over the next decade. The Atomic Industrial Forum (AIF) has projected that installed capacity will about triple by 1990. This implies production of about 4 mbde. AIF data indicate that if all reactors under construction and ordered could be completed by 1990, installed capacity would increase 3.5 times, implying production of about 4.5 mbde.[2] Figures from the current estimate of the National Electric Reliability Council show that nuclear power generation will increase by a factor of about 3.4 by 1989.[3]

Nuclear energy can grow substantially over the next decade with due attention to the problems of (1) availability of enrichment services, (2) spent fuel storage, (3) licensing and (4) safety.

Enrichment Services

Continued development of nuclear power over the next decade depends upon having an adequate supply of nuclear fuel. That, in turn, depends on having adequate uranium reserves, on having mines and mills to extract the uranium and on having enrichment facilities to concentrate the product for use in operating reactors.

According to the Department of Energy, the United States has 690,000 tons of known uranium "concentrate" reserves and considerable amounts of probable reserves.[4]

Of these reserves, the Nuclear Energy Policy Study Group (a group sponsored by the Ford Foundation, which made its report in 1977) stated that "uranium resources will probably be adequate well into the next century."[5]

Annual production of uranium, however, must grow to operate existing and planned power plants. Uranium concentrate production in 1978 amounted to 18,500 tons. Mills under construction or in the planning stage would double current production capacity. These increases will probably take place within the next five years. Thus, it appears that production capacity will be adequate during the decade.

A potential problem is the future availability of enrichment services to support domestic nuclear plants. At present, the government is the sole domestic provider of this service, with three plants. According to current plans, the processing capacity of these plants would be increased nearly 60 percent by 1981. Another plant is planned for the end of 1988.

Enrichment capacity is considered to be adequate, therefore, for the next 4 to 6 years, but plans to add to the enrichment capacity must be followed through if 1990 forecasts of nuclear experts are to be valid.

Storage of Spent Fuel

The storage of spent fuel is frequently referred to as an "unsolved" problem, but nuclear experts feel that this description is misleading. "All this means," according to Bernard L. Cohen of the University of Pittsburgh, "is that a method of disposal has not yet been decided upon."[6]

Indeed, Dr. Cohen asserts that nuclear wastes are less dangerous than coal-burning wastes. In an article published in the November 1978 issue of *Commentary*, he observed: "They [nuclear wastes] are far less hazardous to dispose of, whether one thinks in terms of traditionally simple and cheap disposal methods or the more sophisticated methods made possible by present and near future technology."[7]

The technology for the safe disposal of nuclear waste material has been available for some years. In 1976, the

Electric Power Research Institute Journal (EPRI) pointed out that "the technical feasibility of several [waste management] processes has been adequately demonstrated on at least pilot scale for each required step, from the discharge of spent reactor fuel to the disposal of radioactive waste in terminal storage."[8]

If the technology is available, what has caused the delay? According to EPRI, the holdup has been caused by "industry awaiting stable regulation before committing reprocessing facilities, government awaiting more data before setting regulations, and researchers awaiting regulations before being able to produce precisely that data needed."

In the early period of nuclear energy development, it was envisioned that spent fuel would be cooled and then reprocessed for the recovery of residual uranium and newly generated plutonium. In 1977, however, President Carter deferred reprocessing as part of a plan to reduce the proliferation of nuclear weaponry. A task force named the Interagency Review Group on Waste Management was charged with developing recommendations for waste storage and disposal. Meanwhile the delay has given rise to problems of interim storage.

The nation's commercial reactors are currently storing their spent radioactive fuel in on-site underwater tanks. While awaiting government decisions on both interim and permanent storage, utilities have expanded their on-site storage facilities. Planned storage facilities for reactors still in the design stage have been enlarged.

In February 1980, President Carter announced a new storage program, which should help resolve some of the uncertainties in the waste disposal picture.

The new program proposes (1) extensive research and development of a number of geologic formations before selection of a repository site; (2) a government contingency program for storage of spent fuel and disposal of low-level nuclear waste; (3) licensing by the Nuclear Regulatory Commission of any future nuclear waste management facilities; and (4) creation of a State Planning Council to advise the executive branch on nuclear waste policy including interim storage of spent fuel.

The State Planning Council will be made up of 14 members from various states, the secretaries of Energy, Interior and Transportation, and the administrator of the Environmental Protection Agency.

In announcing the program, the president said the United States should be ready to select the site for the first full-scale repository by about 1985 and have it operational by the mid-1990s. In the meantime, the government will provide -- if Congress approves -- temporary away-from-reactor storage capacity for spent fuel that cannot be stored at expanded reactor pools.

If this step is taken, all reactors can continue to operate through the 1980s and beyond.

Licensing

Even before the Three Mile Island accident, licensing of a power reactor had become an extraordinarily lengthy and involved process. A reactor must be licensed twice: first, to begin construction and then to begin operation.

The licensing process involves not only the filing of formal applications by the utility, but also a lengthy review procedure by the NRC. This review includes further studies, conferences with the applicant, and public hearings.

Siting and environmental factors contribute to the delays, in part because federal, state and local agencies often conflict in the many areas they regulate. A company might have to obtain more than 50 permits to comply with federal, state and regional regulations.

As a result, according to the Atomic Industrial Forum (AIF), a utility must now allow 12 to 14 years for bringing a nuclear power plant from initial planning to commercial operation. And with each month's delay in start-up, the AIF adds, $10 million is added to the cost of a typical generating station.[9]

The impact of the cumbersome review process is widely recognized within the government. The 95th Congress considered (and failed to pass) the administration's proposed Nuclear Licensing and Siting Act of 1978, which would have streamlined licensing procedures.

Now, in the aftermath of Three Mile Island, a six-month "pause" in the licensing process has been ordered by the NRC until further safety requirements can be developed. In advance of that deadline, some limited or "restrictive" licenses have been issued, authorizing low-power testing of completed installations. The administration expects that full licensing will be resumed soon.

Nonetheless, Congress still needs to take action to provide for early site review, combined construction permits and operating licenses and limits on the reconsideration of issues previously decided. These actions are important in facing up to America's import problem. Each 1,000 megawatt reactor can save the equivalent of roughly 30,000 barrels of oil a day.

Nuclear Reactor Safety

The nuclear industry has an impressive safety record, paradoxically brought to public attention during the national reaction to Three Mile Island. In more than two decades, the AIF points out, no injury to the public has occurred, and no plant employee has been seriously injured by exposure to radiation.

Scores of studies and investigations were conducted into the causes of the Three Mile Island accident. Most of the studies focus on the element of "human error" and recommend improvements in operational training and procedures.

The NRC has reviewed thousands of recommendations made by the various studies, categorized them and assigned priority to each. An NRC "action plan" will recommend upgrading the training of crews, improving operating procedures, auditing more closely the day-to-day operations and other measures.

The special Presidential Commission on the Accident at Three Mile Island found that the public was exposed to amounts of radiation that were well within safe levels. The commission estimated that the radioactivity released to the 2 million people living within 50 miles of Three Mile Island was approximately 2,000 person-rems.[10] This works out to an average exposure of 1 millirem per person (0.001 of a rem) in the Harrisburg, Pennsylvania, area -- about the same as the additional natural background radiation a resident of Denver receives every week.

John Kemeny, chairman of the commission, told The New Hampshire Times earlier this year:

> I honestly believe the following: If our recommendations are implemented, and I do not mean that every single one has to be followed verbatim, but if the major recommendations and the spirit of our recommendations are followed, I think nuclear power can be made safe enough so that I wouldn't worry about living near a nuclear power plant.[11]

And a report on nuclear power plant safety released in March 1980 by a subcommittee of the House Committee on Science and Technology asserted:

> The radiation exposure to the nearby population resulting from the accident at Three Mile Island was substantially less than that received from a number of other sources.[12]

In other studies, the health risks of nuclear energy have repeatedly been found to be minimal. In 1979, a committee of the National Academy of Sciences concluded that the public's exposure to radiation from the nuclear power program is "trivial" compared to that from medical and natural sources, posing no discernible health threat. And the U.S. Department of Health, Education and Welfare, now the Department of Health and Human Services, after a year-long study, reported that the health impact of occupational radiation exposure in the nuclear industry "is in line with the risks associated with other industries generally regarded as safe."[13]

In the wake of Three Mile Island, the industry has launched an ambitious educational program to make the public aware of the safety of nuclear power and to attempt to respond quickly and accurately with information when mishaps occur.

The industry is also better organized to work on safety problems than it was before the accident. In 1979, the Nuclear Safety Analysis Center, an arm of the Electric Power Research Institute, was formed. In December of that year the Institute of Nuclear Power Operations was established. Both groups are concerned with setting safety standards and procedures and conducting research.

Outlook

One year after the trauma of Three Mile Island, there are reasons to believe that nuclear energy is ready to resume its growth. As the Atomic Industrial Forum points out, there appears to be this growing consensus about nuclear power: "We can do it better, but we cannot do without it."[14]

Polls continue to indicate that the majority of Americans favor the nuclear option -- even after Three Mile Island. In mid-January 1980, a survey by Louis Harris and Associates indicated that 51 percent of the public favored new construction, and 68 percent favored the continuation of power plants now in operation "if they meet strict government safety standards."

Recent developments in Sweden may foreshadow the U.S. climate of acceptance. In Sweden 58 percent of the voters endorsed the use of nuclear energy to the year 2000 -- a remarkable outcome, considering that the country's prime minister and his political party led the anti-nuclear faction.

With a climate of public acceptance and with timely government action on the enrichment, spent fuel storage and licensing problems, by 1990 nuclear power could make a significant contribution to the national goal of cutting imports by as much as 50 percent.

VI. SYNTHETIC FUELS AND RENEWABLE ENERGY

In addition to oil, natural gas, coal and nuclear power, a new generation of energy resources will be contributing to United States energy security in 1990.

These new sources include oil and gas made from coal; oil from shale and tar sands; energy derived from the sun's heat (solar thermal) or from sunlight (photovoltaic); fuels from plant matter (biomass); and electric generation from natural forces such as water, wind, and the earth's heat (geothermal energy). Hydropower currently provides the largest supply of electricity from natural forces although the growth potential of hydropower is less than that of other natural energy sources.

These alternative technologies now provide the nation with oil equivalent of about 1.5 million barrels a day. If our nation adopts a carefully planned but aggressive program of development, production of synthetic fuels and renewable energy in 1990 could supply 3 million to 4 million barrels a day of crude oil equivalent (mbde).

The benefits of synthetic fuels development are far-reaching. Synthetic oil and gas can provide a critically needed supplement to petroleum reserves and, unlike some energy sources, directly replace -- barrel for barrel -- costly and unstable oil imports.

Thus, these new energy sources can contribute substantially to a national goal of reducing imports by as much as 50 percent by 1990.

Resources for Synthetic Fuels

Fortunately, this country possesses the natural and technological resources to make solid progress on synthetic fuels in the next decade. Coal and oil shale are the principal sources of synthetic fuels. The United States has recoverable coal reserves of at least 250 billion tons and recoverable shale oil reserves of about 600 billion barrels. Together these resources amount to more than five times our total recoverable reserves of conventional oil and gas.

Some technologies for these energy sources are already known and considered to be commercially feasible. The basic technology for extracting oil from shale is more than a hundred years old. The conversion of coal into oil or synthetic gas began in the early part of this century. Production of alcohol from biomass is one of the oldest arts, and the use of alcohol as a motor fuel dates back to the invention of the automobile.

Large-scale "synfuel" production, however, will require more than the revival of old technology. Government and industry are spending billions of dollars to research new methods of converting coal, shale and biomass into synthetic fuels. Oil companies particularly are involved in the development of shale oil, coal conversion and solar energy. Their research and development budgets and technical ability are helping to bring supplementary fuels to the point where they will be ready for general use in the late 1980s.

Synthetic Fuels from Coal

Several companies -- both oil and non-oil firms -- are researching methods for producing synthetic gas, methanol (a type of alcohol) and oil from coal. Some synthetic fuel technologies are on the verge of full-scale commercial operation. Several estimates project that by 1990 liquids from coal could supply 0.1 million to 0.3 million barrels a day (mbd) and gas from coal could supply 0.3 mbde to 0.8 mbde -- in all, 0.4 mbde to 1.1 mbde. The significance of this contribution is apparent when one considers that the Iranian shortfall -- which triggered government allocations that caused lengthy gasoline lines -- was in the middle of that range.

The basic technology for producing gas or liquids from coal is well known. Coal is gasified by reaction with oxygen and steam, to produce a mixture of carbon monoxide, hydrogen and other gases. Further steps clean this mixture and use catalysts to convert it into nearly pure methane -- the major ingredient of natural gas. Methanol also can be made from the mixture of gases.

Several methods exist for converting coal into liquid fuel: pyrolysis, solvent refining, direct and indirect liquefaction. Pyrolysis applies heat to pulverized coal in order to break down parts of the coal molecules into liquids and gases. In solvent refining, coal is pulverized and mixed with an oil-type solvent. This process converts coal into a liquid heavier than crude oil. Direct liquefaction uses a catalyst to add hydrogen to either the liquid product from solvent refining or raw coal. Through indirect liquefaction coal first is gasified with steam and then catalyzed to produce a liquid fuel.

Oil companies have been active in research on coal conversion. Over the past 20 years, Texaco Inc. has developed a process of coal gasification which several companies have used as the basis of their own methods. Texaco, Southern California Edison and the Electric Power Research Institute plan to build a small "combined cycle" electric power plant that will run on both gasified coal and steam produced during gasification. It could be in operation as early as 1983.

Gulf Oil Corporation has developed a solvent for liquefying coal. Gulf is now trying to build two commercial-size pilot plants to process 6,000 tons of coal a day into roughly 20,000 barrels a day of liquid boiler fuel. Gulf expects these plants to be completed by 1985 and hopes to expand production to 100,000 barrels a day by 1990. Other oil companies also are moving forward on coal conversion pilot projects.

Several non-oil companies are also involved in research and commercial development of fuels made from coal. W. R. Grace Company is planning a full-scale synthetic gasoline plant that could produce 50,000 barrels of gasoline a day by 1990. The Grace project will use technology developed by Mobil Oil Corporation for converting coal-based methanol into gasoline.

A number of smaller coal gasification plants could be in operation by the mid-1980s. The Great Plains coal gasification project, funded by a consortium of gas pipeline companies, could be in operation by 1984. The plant will produce the oil equivalent of 20,000 barrels a day of high-quality gas that can supplement supplies of natural gas. By 1983, Tennessee Eastman, a subsidiary of Eastman Kodak Company, plans to produce 3,000 barrels a day of methanol from coal gas.

These and many other firms are investing in the development of synthetic fuel from coal. While some of the projects are receiving government financial aid, others are entirely commercial ventures -- a sign that synfuels are no longer strictly experimental. As the price of crude oil rises, synthetic fuels are able to compete with conventional oil and gas. Therefore, commercial production of synthetics from coal is likely to increase over the course of the decade.

Shale Oil

The idea of producing oil from shale rock is not new. Shale oil production preceded crude oil production in the United States by several years. In the 1850s, some 53 domestic firms manufactured oil from eastern deposits of shale. Shale is a marlstone that contains a solid substance called "kerogen," which is partially formed oil.

While the basic technology for extracting oil from shale is more than a hundred years old, several oil companies have been investigating new shale oil processes over the past 20 years. Both Occidental Petroleum Corporation -- with Tenneco Inc. as a recently acquired partner -- and the Rio Blanco Oil Shale Company (Gulf and Amoco) are working on "in situ," or underground, methods of distilling oil from shale deposits in Colorado. In these processes a portion of the rock is mined and the rest is fractured and heated underground. The heat releases the oil into wells leading to the surface. The Chevron Oil Company, Atlantic Richfield Company (ARCO), The Oil Shale Corporation (Tosco) and the Union Oil Company of California also are experimenting with new ways to extract oil from shale.

Estimates for total shale oil production throughout the industry suggest that by 1990, oil shale could provide as much as 0.4 mbd to 0.6 mbd. Some projects are already under way. ARCO and Tosco, for example, are halfway through the process of obtaining permits and expect to complete a 50,000-barrel-a-day shale oil plant by 1984. (Exxon has recently agreed to purchase ARCO's interest in this project.) Union Oil has obtained all the necessary construction permits and could be producing 10,000 barrels a day in 1984. The commercial-sized plant designed by Occidental and Tenneco could produce 57,000 barrels a day by 1990. Gulf and Amoco suggest a similar estimate for their plant. Chevron is engaged in a detailed study of shale oil processes and hopes to produce 100,000 barrels a day by 1990.

Always in search of secure and independent sources of fuel, the military also has backed the development of shale oil. The Department of Defense (DOD) has been involved in shale oil production since World War I when the government established a Naval Oil Shale Reserve near Rifle, Colorado. Over the past five years, DOD has tested refined shale oil products as both aviation and ground transportation fuel. Thus, shale oil could play an important role not only as a commercial fuel but also as part of a military fuel reserve.

Tar Sands

Tar sands are sands found in their natural state saturated with heavy oil or tar. Where the sands are shallow, they can be surface mined and the oil recovered with heat. Where the deposits are deep, underground, or "in-situ," methods must be used. One such method is to drill conventional oil wells and inject steam to melt the tar. Another method is to create an underground fire. In both cases, the liquid is then pumped to the surface through heated pipes.

The United States has considerable tar sands deposits -- enough to equal our proved crude oil reserves -- but most of it is difficult to extract. Tar sands have been found in at least 22 states, although most are in Utah. Sohio has experimented with surface mining Utah tar sands. However, more than 90 percent of the deposits are deep enough to require the more difficult and expensive in-situ methods. DOE is planning an undergound recovery field experiment on a Sohio lease near Vernal, Utah.

Canada has much greater deposits of tar sands that can be recovered by the surface mining method and commercial production is already underway.

But in the United States several factors have inhibited the development of tar sands reserves. First, as noted, the United States has much smaller reserves than Canada, and only a small portion of U.S. reserves are recoverable by the surface mining method. Second, the U.S. surface-minable deposits are a different type of tar sands than Canada's, and separation of the oil from the sands would require a different process. Third, the surface mining process faces environmental problems. Finally, the process requires more water than is readily available in the Utah deserts where most U.S. deposits lie. Thus, production of oil from tar sands in the United States is not expected to be significant by 1990.

Biomass Conversion

Biomass is not only one of the oldest but also one of the most accessible sources of energy. Biomass -- literally the amount of living matter -- includes every kind of organic substance that can be turned into fuel. Wood and dry plant or organic waste can be burned directly to generate heat and electricity. All matter that contains starch or sugar can be converted into alcohol. Bacteria can break down organic wastes into methane.

Biomass conversion has unique advantages over most other energy technologies. First, the resources are renewable, unlike coal or petroleum. Second, conversion of municipal and industrial wastes into useful fuels can help solve two of the nation's problems simultaneously. It can both increase energy supplies and help clean up the environment.

The potential of biomass fuel is considerable. Government and industry estimates suggest that by 1990, biomass energy production could supply 0.2 mbde to 0.3 mbde.

Perhaps the most important biomass fuel is alcohol. A variety of alcohols -- including ethanol and methanol -- can be produced from plant and animal matter. Currently, the four major distillers in the United States process ethanol from cornstarch, local waste, cheese whey and wheat. These liquids can supplement petroleum both as an industrial feedstock and as a motor fuel.

Most manufacturers use ethanol derived from petroleum to make chemicals, solvents, detergents and cosmetics. However, before the growth of the petrochemical industry in the 1940s, manufacturers used ethanol made from biomass. As the price of crude oil rises, biomass-based alcohols once again are competing with ethanol made from petroleum. Twenty percent of the alcohol used by industries is made from biomass.

The use of alcohol as a motor fuel dates back to the 1880s when some of the first automobiles ran entirely on alcohol. During the 1920s, farmers began to mix alcohol with gasoline. Gasohol -- a blend of 90 percent gasoline and 10 percent ethanol -- originally was called agrifuel, agrol or alky-gas.

Alcohol is regarded as a significant supplement to gasoline for two reasons. First, efficiently produced alcohol can stretch supplies of motor fuel. Second, alcohol raises the octane rating and, thus, can help refiners produce high-octane unleaded gasoline.

The Energy Security Act, passed in June 1980, sets a goal for production of 60,000 barrels of alcohol a day by the end of 1982. The act also sets a 1990 goal for alcohol production equal to at least 10 percent of estimated 1990 gasoline consumption.

Current total domestic production of biomass-based ethanol is only 6,500 barrels a day. And roughly half of this ethanol is used in industry, leaving only some 3,250 barrels a day to blend with gasoline.

Thus, total gasohol supply is about 32,500 barrels a day or around 1 percent of unleaded gasoline consumption.

Government and private industry are studying methods of boosting alcohol production so that gasohol and other alcohol fuels can contribute significantly to the nation's energy supply. Distillation methods in current use were developed in the 1930s, and there is substantial opportunity for improvement. Currently, the production of alcohol -- from growing the corn to distilling the final product -- uses more energy than it yields. If distillation methods improve, alcohol could be produced in a more energy-efficient way and thus make a greater contribution to our energy supply.

One step is to burn coal, wood or other fuels, rather than petroleum, in alcohol plants. For example, agricultural wastes could be burned as boiler fuel in stills. Scientists also are experimenting with different bacteria that could turn biomass into alcohol more economically than conventional processes.

Other biomass fuels, such as methane gas processed from plant matter and refuse, currently provide a negligible amount of energy. However, biomass gasification is a major area of research because methane produced from biomass could stretch supplies of natural gas. Interest focuses particularly on the conversion of municipal, agricultural and industrial wastes into methane that could readily replace natural gas in boilers and electric generators.

Questions have been raised about developing fuels from crops such as corn and wheat that are also needed for food. However, food and fuel need not compete if non-food crops are converted into energy. For example, some agricultural waste might be economically processed into fuel -- perhaps as much as 0.5 mbde.

Also, Americans throw away at least 150 million tons of garbage each year. A considerable amount of this solid municipal waste could be converted into energy. However, there are now less than 20 trash-to-energy plants in operation in the United States. And less than 1 percent of the nation's municipal waste is now being used to generate energy. In contrast, Sweden, the Netherlands and Denmark convert 40 percent of their municipal waste into energy.[1]

A February 1979 General Accounting Office study estimates that by 1985 the United States could process 18 percent of the nation's garbage into energy. Municipal solid waste "could provide the nation with annual energy savings equivalent to 48 million barrels of oil."[2] And the federal government can help. As the study also notes, "If technologically and economically viable waste-to-energy systems are to be used on an accelerated schedule in the near- and mid-term, a more active role by the federal government is required."[3] With the government's help, the nation could add to its energy supply and at the same time help solve the problem of waste disposal.

Renewable Energy Resources

While biomass conversion represents both a synthetic and a renewable energy source, several other forms of energy are classified more strictly as renewable. Renewable resources are those that derive energy from natural forces such as the earth's heat (geothermal energy), the ocean's tides, rivers, winds and the sun. Geothermal energy, tidal and water power and wind power can generate electricity indirectly by turning turbines or fans. Two processes use solar power: solar thermal -- the use of the sun's heat for heating or cooling; and photovoltaics -- the conversion of sunlight directly into electricity by means of solar cells.

These different forms of renewable energy could provide the oil equivalent of 2 million to 2.5 million barrels a day by 1990. However, because research is particularly intense in the area of photovoltaics, technological breakthroughs during the 1980s could increase the contribution of solar energy beyond present estimates. Nonetheless, this paper does not assume such breakthroughs. If they occur, they will contribute additionally to this country's energy security.

Solar Technologies

Solar thermal systems are the most developed and simplest of the solar technologies. Passive solar systems are architectural designs that take advantage of site and building materials to turn the building into a solar collector. In effect, the passive solar system is a form of conservation.

Active solar thermal systems involve mechanical moving parts. The basic unit is the solar collector -- a panel often made of aluminum, glass, plastic and copper. These panels, fitted to a roof, absorb direct sunlight and transfer heat to a fluid that passes through the collector. The fluid flows through pipes into the building, where it can be used to heat water, warm rooms or even provide heat for industrial processes.

The Department of Energy estimates that in 1979 energy from solar collectors amounted to the equivalent of about 3,800 barrels of oil a day -- less than 0.02 percent of all energy consumed in the United States. DOE based its estimate on the 40 million square feet of known solar collectors currently in use.[4]

If present production trends continue, the quantity of solar collectors is likely to double by 1990. Thus, solar thermal systems alone could contribute the equivalent of 8,000 barrels of oil a day by 1990. Economic factors will determine further increases in the manufacture of solar panels. New technology is required to lower costs and increase the competitiveness of solar thermal units. Also, increased production depends on the availability of aluminum, copper and other materials required to produce panels.

Photovoltaics -- solar cells -- are electrical "semiconductors" made primarily from silicon that convert sunlight into electricity. Although the photovoltaic effect has been observed for almost 40 years, the space program first developed the technology in the 1950s and 1960s. With the advent of the oil embargo and higher prices, research on solar cells by the private sector accelerated in the 1970s.

Because the materials are expensive, commercial production of photovoltaics is far smaller than that of solar thermal systems. At present the government heavily supports the industry through research grants and through purchases. The federal government buys roughly one-third of all the cells produced. However, until costs are reduced, it is unlikely that photovoltaics will attain wide commercial use. Currently, electricity generated from photovoltaic sources is at least ten times more expensive than conventionally generated electricity.

Despite the current impediments to commercialization, both government and industry are increasing their research and development budgets for solar technology. Some oil companies are engaged in photovoltaic development. Exxon Corporation, ARCO and Shell Oil Company produce solar cells commercially while at least five other petroleum companies are conducting research. Exxon's involvement in solar cell research began in 1969. Before the oil embargo, the company's internal research and development funding for solar cell research was larger than that of the federal government. To date, total oil company investment in solar energy -- both thermal and photovoltaics -- is in the area of $100 million.

Since the oil embargo, some 245 firms that market solar thermal systems commercially have sprung up. Most of these companies are relatively small firms with annual sales of less than $100 million each. Exxon and ARCO (through their subsidiaries) are the only oil companies that sell thermal systems commercially.

Hydropower and Geothermal

Hydropower and geothermal energy are two important resources that already contribute significantly to the nation's energy supply. Hydropower -- electricity generated by water as it falls and turns turbines -- supplies 1.4 mbde. The share of conventional hydropower, however, is not likely to increase markedly because sites for new dams are limited.

Geothermal energy is produced from heat deep within the earth. Beneath the earth's surface is molten rock which in some places has pushed up close to the earth's crust. When underground water flows over these hot rockbeds, the water or steam rises through cracks in the earth. Geysers such as Old Faithful in Yellowstone National Park are a visible sign of geothermal energy.

Geothermal resources in the United States are tremendous and could provide 0.3 mbde by 1990. Currently only one geothermal source supplies commercial electric power. For nearly 15 years, Union Oil Company has supplied steam from The Geysers field in northern California to electric utilities. The steam from the field helps generate most of the electricity in the San Francisco area.

Government and industry are experimenting with other forms of geothermal energy. At Raft River, Idaho, and in the Imperial Valley of California, engineers are using warm spring water to boil isobutane which turns an electric turbine. The advantage of the system is that isobutane boils at a lower temperature than water. So, when natural spring water is not hot enough to produce steam, it can still boil the isobutane.

In southeast Arkansas, DOE is experimenting with waste water in a bromine mine to produce electricity. Engineers pump out a hot brine that contains bromine. After removing the bromine from the natural spring, the waste hot water is used to generate electricity.

Before these experimental projects become commercially viable, numerous economic and technical problems must be solved. For example, drilling costs in geothermal zones are twice as high as in onshore oil or gas wells at similar depths. Equipment also needs to be developed that can resist the high temperatures deep within the earth. However, the operation of The Geysers field indicates that some geothermal resources can be commercially developed under existing economic and technical conditions.

In addition to geothermal and hydroelectric power, other natural resources such as wind and tidal forces could become sources of energy in the future. Scientists currently are conducting research on ocean tides, and the aerospace industry is investigating new methods of generating electricity from wind. However, these unconventional sources of energy will not be significant by 1990.

Getting Alternative Fuels to Market -- Problems and Opportunities

The 1980s will be a decade of transition for alternative fuels. Some processes now under study may never become commercially feasible, while others may have only limited application.

Since many synfuels processes have never been in commercial use, their costs, large-scale feasibility and environmental impact are difficult to assess. Because of these uncertainties, the United States could benefit from a carefully planned but flexible program for developing alternative energy resources. Logically, the program would not commit consumers to any one process but would allow the market to select the competing technologies best suited to the public's energy needs.

Energy companies have developed the technological expertise and management skills to help build a new synthetic fuels industry. In most cases, however, the necessary technology is still in the research and development stage or simply is too costly to compete with conventional fuels.

World crude oil prices are $25-$35 a barrel or more. By comparison, most estimates of the equivalent cost of alternative fuels range from $40-$70 for alcohol from biomass; $30-$36 for gas from coal; $30-$50 for liquid fuel from coal; and $30-$50 for shale oil. Shale oil and gas from coal are closest to commercialization in the United States.

Normal marketplace development will depend on how these various costs compare in the future. However, with the future security of U.S. energy supplies at stake, the government can help promote more rapid development. This government assistance should rely on the market system, with the government setting the ground rules and the private sector making management decisions. The program should be simple, encourage widespread participation, encourage efficient design and operation and allow timely construction. Such an approach would assure consumers the greatest amount of synthetic fuel production at the lowest cost.

The private sector should assume full risks and responsibilities for building and operating commercial-scale plants. However, government can play a key role in research and demonstration projects that would not be developed as quickly under normal market conditions. For projects that are close to being commercially competitive, incentives should include accelerated depreciation, energy investment tax credits and production tax credits. Loans, loan guarantees and guaranteed purchase price contracts may be appropriate in some cases as long as the producers still bear a significant share of the project risk.

In addition to providing economic incentives, government can improve the atmosphere of synfuel development by modifying some existing laws that may inhibit the development of these alternative fuels. The same environmental restrictions that frequently delay conventional energy projects will impede the production of synthetic fuels from coal and shale. Under current conditions, the lead times required for most synthetic fuels projects average seven to eight years: two to three years for obtaining permits and five years for construction. If synthetic fuels are to have an impact in the 1980s, government will need to examine and reform the regulatory process, especially with regard to environmental regulations.

Modifications in the Clean Air Act and other laws can cut the red tape and speed the process of review and approval required for synfuels development. For example, the Environmental Protection Agency originally refused permission to the Rio Blanco Oil Shale Company to build a shale oil plant in western Colorado because natural pollutants such as dust caused the air quality of the area to fall short of EPA standards. After monitoring the area for one year, Rio Blanco demonstrated that the EPA standards were unrealistic. EPA reinterpreted its regulations on natural emissions and subsequently granted permission for the project.

In a report released in early 1980, the House Science and Technology Committee's Advisory Panel on Synfuels noted that the demonstration phase of a commercial synthetic fuels program can be operated in a manner "consistent with present substantive and procedural environmental laws."[5] However, the panel also suggests that wide-scale commercial development of synfuels "may require some modification in those laws to meet particular problems."[6]

Questions have been raised about possible unknown effects of synthetic fuel processes on human health. Government and industry are closely monitoring experimental work on coal conversion, shale oil and biomass. Much still remains to be learned before various synfuels projects reach the commercial stage. But some of the data now available suggest that synfuels development may pose fewer environmental hazards than anticipated.

The Federal Interagency Committee on the Health and Environmental Effects of Energy Technologies recently reported that synthetic oil and gas from coal "may be environmentally compatible and clean burning."[7] The committee also stated that, while coal conversion may create health risks, "the actual hazards may be no higher than those from conventional coal or petroleum refining operations."[8]

Many uncertainties are bound to arise in the commercial development of new energy technologies. But the benefits to the public of developing alternative energy resources dictate that we press prudently ahead. In the short term, synthetic fuels can help reduce dependence on foreign oil. In the long term, they could provide an abundant and secure supply of energy.

VII. ENERGY CONSERVATION

The United States has another important energy resource -- conservation. More efficient use of energy can help substantially reduce oil imports.

Americans are using less energy this year. Energy consumption is well below estimates projected only a few years ago for 1980. These reductions have occurred because of rising energy costs, intermittent cutbacks and embargoes of petroleum imports and mandated fuel savings, such as the national 55 mile-per-hour speed limit and automobile gasoline mileage standards.

As a result, analysts now believe the country will need 10 million barrels a day of oil equivalent (mbde) less in 1990 than was projected for 1990 only a few years ago.

These reductions in energy consumption growth trends have prompted suggestions that conservation, by itself, can solve the nation's energy problem. Recent public opinion polls, in fact, show that a majority of Americans believe that energy conservation is the single step necessary to reduce U.S. dependence on imported oil.[1]

Such views are unrealistic. The potential for saving energy is not unlimited. And, even some experts who advocate strict conservation measures anticipate that the United States will need 15 to 20 percent more energy in 1990 than we are using today to meet the needs of a growing population and economy.

But while it is not the total answer, conservation will continue to be an important response to the nation's energy problems. Since it has the effect of enhancing domestic energy security -- much like a new oil recovery process or opening a new coal mine -- conservation can be a key element in a national goal of reducing U.S. oil imports. Government, academic and private research studies emphasize its importance in meeting future energy requirements. Indeed, conservation has become a major objective of national policy, advocated from the highest levels of government.

Of all the options available to Americans, conservation can be the cheapest and fastest means of decreasing the nation's dependence on foreign oil. Through sound conservation policies and practices, the nation can quickly and significantly reduce the rate at which energy consumption rises, while still permitting an overall economic growth rate consistent with public expectations. And, through conservation, all Americans can participate directly in improving the country's energy position.

Effects of Conservation

Americans already are using energy more carefully. During the last seven years, as the cost of energy has gone up, the rate of growth in consumption has dramatically declined.

During the post-war years, and particularly the years just prior to the 1973-1974 embargo, the average annual growth rate for energy in the United States was about 3 percent.

Immediately following the embargo and subsequent quadrupling of world oil prices, total U.S. energy consumption actually declined -- dropping from 35.2 million barrels of oil equivalent per day in 1973 to 34.4 mbde in 1974, and to 33.4 mbde in 1975.

As world prices stabilized in the mid-1970s and the U.S. economy recovered from the post-embargo recession, energy consumption rose again in 1976, climbing to 35.1 mbde -- the 1973 level -- and amounting to a 5.4 percent growth rate over the previous year.

Since that one-year spurt, however, U.S. energy growth rates have moderated, averaging about 2.5 percent a year between 1976 and 1978, and changing little from 1978 to 1979.[2] The 1979 leveling off came mainly in response to the doubling of world oil prices as a result of OPEC actions. Although consumption is expected to rise again, forecasts now show the average growth rate for the 1980s will be less than increases seen over the last five years.

Over the last five years, projections for total U.S. energy consumption in 1990 have dropped by about 20 percent because of higher prices and conservation efforts. The 1979-1980 forecasts for 1990 show an increase in consumption from today of about 5 million to 10 million barrels a day of oil equivalent, to a total of about 42 million to 47 million barrels a day. This compares with 1990 consumption estimates of 50 million to 55 million barrels a day made just two to five years earlier.

With regard to petroleum energy, specifically, in 1979 the United States consumed 413,000 barrels of oil a day less than in 1978, a drop of 2.2 percent. Gasoline consumption alone decreased by 5.2 percent in 1979, and for the first few months of 1980 it ran from 8 percent to 12 percent below year-earlier levels.

In 1979, too, the United States was the only major industrial nation of the free world to show a decrease in consumption of oil. According to the Central Intelligence Agency, oil consumption for six other major industrial countries increased.[3] The increases ranged from 0.5 percent for the United Kingdom and 1.1 percent for Japan to 3.6 percent for Italy and 4.4 percent for Canada.

With more fuel-efficient new cars, with consumers turning more and more to smaller vehicles, with business introducing more efficient equipment, with homeowners installing more insulation and with other deliberate cutbacks, evidence is piling up to demonstrate the potential savings of conservation.

However, there are limits to the amount of energy that can be saved. Even the most efficient car will need fuel. Energy requirements in homes and industries could be cut by as much as half, but homes and industries will still need natural gas, oil, electricity and coal. Thus serious conservation will be a necessary part of the effort to make the United States more energy sufficient. But, the vigorous development of domestic energy resources is also essential.

Public Information Program

Both government and the private sector have expanded efforts to encourage public conservation of energy.

Many corporations, trade associations and private research groups regularly promote conservation measures. These range from urging motorists to save gasoline to special publications and seminars that offer tips on car maintenance, home insulation and good driving.

A number of companies also are subsidizing van pooling and mass transit commuting. Some oil companies and utilities provide residential energy audits for their customers to show where additional savings can be achieved through better insulation, heating systems or energy use practices. One company is offering, at cost, timing mechanisms that automatically raise and lower home thermostat settings. Others are underwriting an information-gathering program on energy conservation by the Carnegie Institute. And, one oil company has loaned a department manager to the city of Los Angeles to help form a car pool information program.

Information programs are the focus of several pieces of federal legislation, as well. Two bills being considered by Congress require individual states receiving federal financial assistance under the legislation to provide energy extension service activities, including advice and assistance on energy conservation matters. States are asked to consider programs to promote the use of car pools, van pools and public transportation.

Also, information can help consumers who are not able to determine on their own the energy efficiencies of goods they purchase, such as houses and appliances. To this end, mandatory efficiency labeling has been proposed. Although some producers already provide such information, there are differences of opinion on the wisdom of mandating labeling.

Where the Biggest Energy Savings Can Be Made

Two big areas for energy conservation are homes and cars.

Nearly 30 percent of U.S. energy demand occurs in the form of direct consumption for personal use, namely household and transportation fuels. Home heating accounts for about 10.5 percent of total U.S. energy consumption, while private passenger cars account for about 13 percent.

The Carter administration estimates that well over 500,000 barrels of oil per day could be saved through residential conservation alone over the next decade.[4]

Residential

The number of homes will be expanding over the next decade or two, so conservation measures in new and existing homes will be important. A study made in 1977 projected that the number of homes in the United States will grow from 70 million in 1975 to 114 million by the year 2000.[5] The study also projected a drop in occupancy rates per dwelling from 3 persons to 2.3, with homes having more floor space.[6] Although later studies may show some downward adjustments in these projections because of current economic conditions and rising energy costs, the preference of Americans for single-family homes is expected to continue to boost housing construction.

Heating accounts for 50 percent of all residential energy use, so significant energy savings could be achieved through:

- improvements in housing shells (in the form of better insulation, weather stripping, caulking and storm windows);

- more efficient heating systems in new housing and replacement of older units (for example, heat pumps to replace conventional electrical resistance heat systems); and,

- improved house designs and layouts (for example, double-wall construction, the addition of vestibules, and windows facing south).

The staff of Resources for the Future (RFF), an independent organization for research and education in the development, conservation and use of natural resources, has examined home energy uses. RFF found that it would be cost effective to reduce the energy consumption of many homes by 45 percent.[7] Further, it projected that energy requirements for home heating will decline slightly by the year 2000 despite a more-than-50 percent increase in the number of dwelling units. Improved residential heating conservation was estimated to have a capital cost of just $500 per dwelling unit (in 1975 dollars).[8] Thus, aggregate energy consumption for space heating could decrease by almost 50 percent by the year 2000 if a more aggressive conservation program is pursued and if energy prices rise at faster rates than projected in earlier estimates.

RFF also noted that major energy savings could be achieved to reduce total household energy consumption in the years ahead with little or no effect on individual lifestyles or projected home construction.[9]

Automotive

Conservation obviously is important in the transportation sector. Energy used in transportation -- almost entirely from oil -- has accounted for about one-fourth of the nation's energy consumption in recent years. And private passenger cars account for about 50 percent of the transportation sector's energy demand.

Significant energy savings are expected by the year 2000 due to government-imposed fleet fuel standards and to higher fuel prices. Current standards require the fuel economy of new automobiles sold in 1985 to average 27.5 miles per gallon. Under these standards, total fuel consumption is expected to fall 31 percent by the year 2000 -- below levels prevailing in 1976. Further, several federal agencies are proposing higher mileage standards which, if imposed by government, would further reduce fuel consumption.[10]

Additional reductions could be achieved if the public voluntarily cuts back automobile use even further and makes wider use of commuter van pools.

Also, various controversial federally-mandated programs have been proposed, such as rationing gasoline, closing service stations on weekends, restricting driving on certain days or raising excise taxes on fuel.

Compared with other industrialized countries in Western Europe and Japan, the United States uses energy for automotive transport more intensively. The causes are easily identified. Gasoline prices have been significantly higher abroad, largely because foreign governments have imposed stiff taxes on gasoline to raise revenues. In addition, public transportation in other countries is more developed and widely used because fares are subsidized, because the cities are denser, and because many of the countries also impose high taxes on cars, thus discouraging automobile ownership.

Other Energy Savings

Substantial energy savings could be achieved in certain industrial processes, as well. Energy for industrial process steam, for instance, accounts for about 14 percent of total U.S. energy consumption.[11] Fuel consumption in this category could be reduced by as much as 30 percent by employing cogeneration -- the combined production of electrical or mechanical energy and heat.[12]

Cogeneration in the United States has been inhibited by the high initial investment usually required and the failure of state public service commissions to adopt electric rate structures that would encourage the practice.

Energy Conservation by Industry

Since 1972 -- the last full year before the oil embargo -- business and industry in the United States have become much more energy efficient. Production has increased nearly 27 percent,[13] while energy use has increased only 7 percent.[14] Savings such as these are part of a recognition that energy conservation results in benefits both to stockholders and to the general public.

Conservation by business and industry has been, in large part, a free market response to increased prices. In 1974, for example, U.S. petroleum refiners voluntarily decided to try to cut energy use in refineries by 15 percent by 1980. By the end of 1977, the refiners already had surpassed the 15 percent mark, which in the meantime they had raised to a new goal of 20 percent. By the end of 1978, these refiners had achieved a reduction of 19.6 percent.[15]

These savings have been achieved through better training, better "housekeeping" features and capital investment in equipment such as improved instrumentation, heat exchangers and pre-heaters.

In all, the 10 industries which were required to report energy efficiency improvement progress to the government for 1978 exceeded the target of 13 percent originally set for 1980, with these industries reaching a level of 14 percent composite improvement in 1978.[16]

Barriers to Energy Conservation

While everyone agrees that conservation has great potential for reducing the growth of energy consumption, there are various barriers to increased investment for conservation.

The building industry finds innovation difficult because of local building and zoning code specifications and because of the diversity of the industry. Cogeneration is hindered because of costs and rate structures. Homeowners, although enthusiastic about conservation, put off purchases for insulation or other goods because they cannot afford them or believe government eventually will pay part of the costs, or both. Finally, many low-income consumers have difficulty in financing conservation investments.

Toward a National Conservation Policy

Some public policy groups suggest that government impose stricter measures -- beyond the free market -- to spur conservation. These include subsidies that go far beyond those now in effect, new regulations and new taxes.

On the other hand, Stuart E. Eizenstat, assistant to the president for domestic affairs and policy, has stated, "True conservation cannot come solely from massive government subsidies -- as important as they are. It can only come when consumers of energy recognize that they can save money through better conservation practices."[17]

Under normal circumstances, it would be preferable to let individuals freely determine the appropriate amount of energy conservation. However, oil imports may pose certain national security risks, and hence some government incentives may be appropriate to encourage energy conservation beyond what would occur in free markets. The provision of government incentives can be justified only if the benefits to the public in terms of enhanced security outweigh the costs.

An effective conservation incentive program should be simple, market-oriented and encourage widespread participation, while minimizing the costs to the taxpayers.

Conclusion

Energy conservation can contribute substantially to the goal of reducing the nation's dependence on imported oil by as much as 50 percent by 1990. Evidence from recent years suggests that voluntary energy conservation in response to price changes has been significant.

But if conservation is to become a way of life for Americans, realistic and reasonable policies will have to be adopted.

Energy price decontrol is a step in the right direction. Decontrol will itself induce energy conservation, and it also will encourage the production of domestic substitutes for imported oil. The country is using less energy today because of higher prices and conservation.

Considering the impact of conservation, the consensus of "best case" forecasts is that, in 1990, U.S. energy requirements will be in the neighborhood of 42 million to 46 million barrels a day of oil equivalent -- 10 million barrels a day below earlier projections.

Through additional efforts, conservation probably could contribute still further toward a national goal of cutting imports. How much more conservation contributes will depend on voluntary savings of energy by Americans and the policies adopted by government to encourage additional conservation.

VIII. ENERGY AND THE ENVIRONMENT

If we increase energy production, as we can, we must do so with careful regard for the environment. This can be done.

The last 10 to 15 years saw a justifiable increase in public concern over environmental pollution. With public concern heightened, the federal government responded with a series of major environmental laws during the decade of the 1970s.

As a result, the United States has made rapid progress toward meeting the high environmental goals established by government. Further environmental progress can be made as we pursue the goal of reducing our heavy dependence on foreign oil by increasing the development of our own vast energy resources.

Significant Environmental Progress Has Been Made

The progress made in restoring the nation's environment is illustrated by the considerable improvement in air quality over the last decade.

The dramatic advances include:

- a 30 percent decline in average sulfur dioxide levels in urban areas since 1970;
- a 40 percent decrease in emissions of particulate matter (dust, soot, etc.) nationwide over the past 10 years;
- an 80 percent to 90 percent drop in carbon monoxide and hydrocarbon exhaust emissions for 1978 and later model automobiles, compared with cars manufactured a decade earlier; and
- an 85 percent reduction in hydrocarbon evaporation losses from gasoline tanks and carburetors of automobiles over the past decade.

The petroleum industry itself, in responding to environmental concerns and regulations, has invested more than $15 billion in air, water and land environmental improvements since 1970. Examples of the industry's progress include:

- reductions in emissions and odors at refineries through installation and operation of new equipment and procedures;
- manufacture and widespread availability of unleaded gasoline;
- development of low-sulfur heating oil and residual fuel oil;
- creation of systems that return water (usually from a river for use in refining cooling and processing) to the river in the same state of purity;
- establishment of two multimillion dollar compensation funds to reimburse victims of oil spills from tankers; and
- development of sophisticated equipment and procedures to prevent or control blowouts from offshore petroleum operations.

The environmental laws adopted during the 1970s -- together with the efforts of concerned individuals and groups in both public and private life -- have moved us well along toward the goal of a cleaner environment . . . a goal identified more than a decade ago.

Further Environmental Progress Can Be Achieved

Now, in this new decade, we face the crisis of heavy dependence on insecure foreign oil. The nation's essential new goal must be to reduce that dependence, especially from the unstable areas of the Middle East and Africa. This goal can be accomplished through development of our own vast energy resources.

For at least the next 10 years, the nation's conventional energy resources -- oil, natural gas, coal and nuclear power -- will be called on to furnish most of the nation's increasing energy requirements. Later, nonconventional sources will begin to play a substantial role in meeting the energy needs of American consumers. All of these vast resources can be developed, and Americans can benefit from both secure domestic energy supplies and further improvements in the quality of the environment.

In short, environmental and energy needs and goals are compatible and can be balanced. However, to achieve this balance, energy development must be considered as important as environmental improvement.

With a wide array of existing and emerging energy resources, the United States can be flexible in pursuing tomorrow's energy goals. With substantial environmental progress already achieved, we need to be more flexible today in pursuing the nation's remaining environmental goals.

However, up to now, rigidity rather than flexibility has often characterized environmental laws and regulations.

This rigidity is reflected in more than a dozen major environmental laws enacted in the 1970s. The intent and goals of these laws have often been good. But, the restrictions and red tape built into the laws and their regulations have worked against development of much of the nation's vast energy resources.

The laws include the following: National Environmental Policy Act and Clean Air Act Amendments (adopted in 1970); Alaska Native Claims Settlement Act (1971); Coastal Zone Management Act, Marine Protection Research and Sanctuaries Act, and Water Pollution Control Act Amendments (1972); Endangered Species Act (1973); Deepwater Ports Act and Safe Drinking Water Act (1974); Coal Leasing Act Amendments, Federal Land Policy and Management Act, and Resource Conservation and Recovery Act (1976); Surface Mining Control and Reclamation Act and further amendments to the Clean Air Act and Water Pollution Control Act (1977); and Outer Continental Shelf Lands Act Amendments (1978).

For example, the Surface Mining Control and Reclamation Act has led to regulations from the Department of the Interior that will prohibit production of some of the most economically recoverable coal deposits. The regulations could also cause mining in some areas of the country to be shut down. As another example, the Outer Continental Shelf Lands Act Amendments, with 40 new or revised sets of regulations, will significantly delay leasing, exploration and production of oil and natural gas discovered on the Outer Continental Shelf (OCS). The Amendments may also discourage some companies from exploring for new petroleum reserves on the OCS.

The goals and intent of these environmental laws need not be altered. Instead, the nation needs as soon as possible more flexible government actions relating to the regulation and administration of these laws.

The impact of these laws and regulations to date was succinctly summarized in a recent article in Harper's magazine (April 1980). The article, by Philip K. Verleger, Jr., senior research scholar at the Yale School of Organization and Management, reported that three recent studies point out that "the United States should consume less (conserve) and produce more fuels in order to solve its energy difficulties. The authors [of these studies] suggest, however, that current laws and regulations preclude a solution."[1]

The Harper's article goes on to cite a finding in one of the studies (Energy: The Next Twenty Years, sponsored by the Ford Foundation) that "the present form of U.S. legislation is far more restrictive than is necessary to achieve the nation's environmental goals."[2]

The same study goes on to say, "Air pollution control is probably the only area of public policy that uses brute force regulation to try and solve a resource allocation problem of such magnitude and complexity." And the study adds, "The debates over setting the ambient air quality standards, the deadlines, and the technological rules and then changing them every few years result in delay, cost, and uncertainty while doing little to improve air quality."[3]

Verleger sums up the arguments advanced in the Ford study by stating that present regulations "have not worked well; permit existing sources to pollute at excessive rates while newer, potentially cleaner facilities cannot be constructed; and discourage research and development for better control technology, especially for existing sources."[4]

The need to formulate a balanced and flexible energy-environmental policy is reflected in an April 1980 statement by Stuart E. Eizenstat, assistant to the president for domestic affairs and policy. Mr. Eizenstat said:

> We cannot return to the days when the environmental impacts of our actions were ignored, but neither can we ignore the tough compromises and trade-offs which will need to be made in this new decade to meet these sometimes contradictory challenges.[5]

With more reasonable and flexible environmental laws and rules, the United States can:

- produce additional supplies of domestic oil and gas, onshore and offshore;
- step up the mining and use of abundant coal supplies;

- develop the vast deposits of oil shale in the western states; and

- convert some of the huge reserves of coal into synthetic oil and gas.

With more rapid development of available domestic energy, we can meet our energy goal of cutting imports by as much as one-half by 1990 without reversing the environmental progress made in many urban and other areas.

Environmental and Energy Needs Can Be Balanced

The American public can have the additional domestic energy it needs -- from both conventional and nonconventional sources -- if government policies recognize the importance of achieving both the nation's energy and environmental goals.

Several avenues are available to achieve this balance.

1. Government can review and adjust environmental standards and rules that have had, up to now, unforeseen, unintended and unnecessary adverse effects on energy development and economic stability.

This is especially important in the case of the Clean Air Act. That act, as it now stands, subjects every square inch of the country to extremely limited-growth and, in some instances, no-growth policies. Unless the act is modified soon by Congress, it will retard the development of various forms of energy -- from coal to geothermal power -- as well as industrial expansion.

The act should be modified by Congress:

- to ensure that the national air quality standards are based on sound scientific, medical and economic data. (The current standards are based on studies done as long as a decade ago. They do not reflect the extensive research, done over the past 10 years, that provides a wealth of new information about the sources, extent and effects of air pollutants.)

- to permit development of the promising energy potential that lies in those areas of the country that now meet the national standards. (Right now, the "prevention of significant deterioration" review requirements of the act are delaying energy development and industrial expansion projects in these areas. The requirements could cause some of these projects to be cancelled by companies.)

• to allow companies to build new facilities in those areas of the country that have not yet attained the air quality standards. (Currently, a company wishing to build a new facility in such an area must first find ways to eliminate more air pollution than the proposed new facility would emit. This is known as the "offset" policy. If the company does not have an existing facility of its own in that area from which to obtain the needed emissions offset, it has three choices. It must pay for controls on another company's facility, buy facilities such as dry cleaning plants outright in order to close them down or abandon its own project.)

• to provide for realistic schedules and deadlines for achieving the national air quality standards. (The present rigid deadlines require the states to make sure that the standards are attained by specific dates. In most cases, the deadline is 1982. If these deadlines are not met, the federal government could cut off state funds and could freeze energy development and industrial expansion in states that fail to meet the standards.)

We can and should modify the Clean Air Act and other environmental laws -- retaining the beneficial elements and eliminating the undesirable effects. These changes will provide the flexibility needed to move forward to achieve the nation's environmental and energy goals.

2. Government can take steps to modify those environmental regulations and policies that unnecessarily prevent or delay development of energy facilities or supplies.

With respect to Clean Air Act rules alone, currently a proposed new energy facility -- before it can move forward -- must survive up to 13 actions, 37 determinations and 24 opportunities for denial by one branch of government or another for up to 17 different air pollutants.

The same kinds of rules and roadblocks are built into the regulations growing out of other environmental laws.

The intent of these rules can be good. But delay does not necessarily result in better decisions. Delay can, on the other hand, lead to business decisions to withdraw an application or to abandon a proposed project. This, in fact, has occurred.

For example:

-- Sohio marine terminal and pipeline: Cancelled. This pipeline was intended to carry 500,000 barrels a day of Alaskan crude oil from Long Beach, California, to Midland, Texas, and then to the Midwest and the East. The project was cancelled after it became uneconomic following five years of delay in obtaining the required federal, state and local air quality permits.

-- Kaiparowits power project: Abandoned. This plant, to be built at Kaiparowits, Utah, was to provide 3,000,000 kilowatts of electricity to areas of the Southwest. It became uneconomic and the project was abandoned after costs increased sixfold as a result of 13 years of delay in securing the required air, water and other permits.

In other instances, the cumbersome and lengthy nature of the regulatory review mechanism has kept a number of energy projects in "deep freeze" for years.

For example:

-- Santa Ynez (California) offshore project: Delayed seven years. The Santa Ynez unit was acquired by Exxon at a federal government lease sale in early 1968. Exploratory testing indicates that this unit could produce 27,000 barrels of oil and 30 million cubic feet of natural gas per day. The start-up of production has been delayed for seven years by federal, state and local regulatory obstacles. During that time, there have been three major environmental impact studies, 21 major public hearings, 10 major governmental approvals, 51 consultant studies and 12 lawsuits. The regulatory hurdles were finally cleared on November 30, 1979, so that construction of the offshore treating facilities could resume. This was more than 10 years after the initial discovery and more than eight years after the company submitted to the government its plans for initial development in anticipation of government approval in about six months.

-- Hampton Roads refinery: Delayed nine years. This refinery, to be sited at Portsmouth, Virginia, would process 170,000 barrels of crude oil a day. The project was held up for nine years primarily because of delays in obtaining approvals and permits from various government agencies. The final government requirement was met on January 25, 1980, although there may still be some legal obstacles.

3. The government can encourage the use of new environmental concepts.

One such approach is called the "bubble concept." It would allow an owner or operator to increase emissions from some sources within an existing plant if he reduces them at other sources within the same plant. The objective is to make sure that the total amount of emissions is within clean air requirements. This concept could provide a measure of the kind of flexibility needed if the United States is to increase domestic energy development and reduce oil imports.

A second new approach would be to establish a "banking" procedure for "offset" credits. As noted earlier, a company wishing to build a new facility or to modify an existing facility is allowed to proceed with its plan only if it eliminates more air pollution than its proposed new or modernized facility would emit.

The "banking" concept would permit companies that reduce emissions from their existing facilities to receive "credit" for doing so. This would encourage companies to make _early_ reductions in emissions or to shut down a polluting facility even if the companies did not propose, at the time, to build a new facility. Companies would receive offset credits which they could either "bank" for possible later use or trade in a free-market system.

Flexible Environmental Policies Can Lead to More U.S. Energy

Prompt government actions to make the nation's environmental policies more flexible can lead to new supplies of most forms of domestic energy.

1. _Oil and Natural Gas Resources_ -- Large petroleum resources can be developed both through new discoveries in new areas and through increased production from existing fields.

Reasonable environmental rules can, for example, speed up the development of oil and natural gas on the Outer Continental Shelf. Right now, approximately 1 million barrels of oil and 14.5 billion cubic feet of natural gas are being produced each day, on average, in an environmentally sound way from offshore wells. We can recover more of that oil and gas through more flexible environmental regulations that expand the leasing of the Outer Continental Shelf beyond the 2 percent that is now under lease.

In addition, billions of barrels of what is called "heavy oil" in known fields could be produced in a relatively short time, primarily in California. Government action could permit oil companies to use properly controlled steam generators. With steam, the heavy oil can be thinned and extracted in an environmentally safe way.

Action by the California Air Resources Board early in 1980 provides some hope that some of this additional heavy oil may be produced from the Kern County, California, field. Estimates of heavy oil in Kern County put the amount at 5 billion or more barrels -- equivalent to nearly two years of oil imports at 1979 levels. The new rules adopted by the California state agency will, according to a spokesman for Standard Oil Company of California, provide "for the protection of air quality and the critical expansion of heavy oil production."[6]

The new rules, in brief, are based on a more reasonable offset policy, which is not quite as severe as it has been for the county. With this more flexible approach, additional steam generators will be allowed to be installed if the operator eliminates air emissions elsewhere in the Kern County oil fields. The offset can come from the operations of the same company or from another company or operator. In some cases, however, it may still be difficult for some operators to add new generators or to "retrofit" existing generators because of a lack of offset opportunities. In addition, if there should be a reduction in the county's air quality, more restrictive requirements could be imposed at a later date.

Nevertheless, it has been estimated that within months after the permits are granted, production increases from the Kern County area could reach 90,000 barrels a day.

Elsewhere in California, additional thousands of barrels of heavy oil could also be produced quickly if balanced energy-environmental regulations were adopted.

Oil and natural gas can also be produced in an environmentally acceptable manner in areas where air quality standards are presently being achieved (the attainment areas). In these areas, provisions of the Clean Air Act have established a policy of "prevention of significant deterioration." The increases in emissions permitted in these areas are so small that they impede and in some instances prevent any kind of energy or industrial development. So-called "buffer zones" around these areas make oil and gas development even more difficult to carry out.

Here is a typical "Catch-22" example of the restrictive provisions of the Clean Air Act. A potential gas reserve of substantial size has been found in southwestern Wyoming, an attainment area. Under the act, it will take some 12 to 18 months to obtain construction permits for production facilities. Then, if the facilities can be built, the strict pollution limits will likely force the company into one of two undesirable choices. One choice would be to construct a recovery plant to control sulfur emissions to the required 99.7

percent level. Meeting this level with a sulfur plant means restricting production to about one-third of the maximum efficient rate. The second choice would be to use a different process that allows production at the maximum efficient rate, but uses more energy to conform to the "prevention of significant deterioration" requirements of the Clean Air Act.

In short, under the first choice, less gas will be produced. Under the second choice, the energy wasted would be enough to heat more than 6,500 homes in the Rocky Mountain area.

The Clean Air Act restrictions can be made less severe; the cumbersome permitting process mandated by the act can be streamlined. With these modifications, the search for new oil and gas reserves in these areas could move forward.

2. Coal Resources -- The production and use of coal can be increased without endangering public health or creating new health risks.

In March 1980, the President's Commission on Coal concluded that substituting coal for imported oil can be done economically and without damaging the environment. Specifically, the commission said, "A program of replacing oil with coal in compliance with the Clean Air Act will not increase emissions of environmental and health-related pollutants and will cause, at most, minimal increases in atmospheric concentrations of carbon dioxide."[7]

Some environmental laws may need to be modified if coal is to make a greater contribution. For the most part, however, the technology for controlling emissions from coal-burning facilities exists.

This technology assures that use of coal will cause no general degradation of the nation's air. There is no ironclad assurance that using more coal will not -- somewhere, sometime -- increase the concentration of one or more air pollutants. Nevertheless, the fact is that each of the primary national ambient air quality standards established by the government incorporates a margin of safety to protect human health.

3. Oil Shale/Other Synfuels -- United States deposits of oil shale contain more oil than all the crude oil reserves of the Middle East. The technology exists to separate refinable oil from shale. The technology also exists to convert much of the huge coal reserves into synthetic oil and gas. This technology is now being tested in pilot plant programs. These new energy resources can, in time, be delivered to consumers if environmental regulations are made more flexible.

Right now, because of environmental considerations, the government review and approval system is delaying the development of synfuels. For example:

• One company had to obtain more than 120 permits and approvals before it could begin experimental work on developing an oil shale site in Colorado. Compiling the environmental data required to obtain these approvals took more than two years. Moreover, these approvals apply only for the experimental production phase of the work. Before the company can begin commercial output -- which it estimates would reach 76,000 barrels a day -- the company will have to go through the whole process again.

• Another company tried to develop oil shale in western Colorado so as to produce 47,000 barrels of refined shale oil a day. The company began providing the results of its environmental analysis to the Bureau of Land Management in 1973, for use in preparing an Environmental Impact Statement. The bureau's statement was not completed and submitted to the Council on Environmental Quality until mid-1977. Nearly two years later (April 1979), the Department of the Interior's Solicitor's Office issued a memorandum that questioned specific aspects of the statement. During the interim, the company had managed to obtain more than 50 permits from the state of Colorado. However, still pending are some 35 permits or approvals from state and local governments -- plus a number of required federal permits and approvals.

Similar restrictions and delays are holding up development of other oil shale and coal conversion projects. Meanwhile, these huge energy resources lie dormant in the ground.

Summing Up

Environmental protection need not mean energy stagnation. And energy development need not mean environmental degradation. We can have both environmental protection and energy development. To achieve this balance, we need to make our environmental laws and regulations more realistic.

Today, these laws and regulations unnecessarily restrict energy development in two ways:

1. Specific laws such as the Clean Air Act can lead to regulations which require pollution levels that are sometimes so low that no one -- including the government -- can monitor them accurately with current field equipment. Even plants with

the best modern pollution controls often cannot meet these standards.

2. The overall maze of red tape, required permits and legal tie-ups have blocked or delayed energy projects of nearly every kind. The energy outlook can be improved by untangling this maze. The key is realistic environmental trade-offs.

The basic question is not whether environmental regulation is necessary but how to regulate wisely. Wise regulation must include simplified permits, cost/benefit analyses, review of existing standards, flexible rules, innovative pollution control concepts and better coordination of federal, state and local regulation.

These more realistic environmental regulations will allow both environmental progress and energy production. And, if initiated soon, they will help cut oil imports by as much as one-half by 1990.

IX. FEDERAL LANDS -- THE NEED FOR ACCESS

Achieving a national energy security goal will depend not only on reasonable environmental regulations, but also on gaining access to areas with the greatest energy potential.

The United States occupies a fortunate position in a world that is hungry for energy. Nature placed vast resources within our borders and beneath our offshore waters.

Americans have used some of those resources to achieve a high level of technology and scientific progress and a standard of living that is unmatched anywhere in the world. However, this country has not come anywhere close to developing its full energy potential.

Many of the high-potential areas for producing oil, natural gas, coal, oil shale, tar sands, geothermal energy and other important minerals are believed to lie within the onshore and offshore areas controlled by the federal government. Hundreds of millions of those acres have never been made available for exploration.

What the Government Controls

The federal government owns 775 million acres -- one-third of the total land area of the United States -- and retains control over the subsurface mineral rights of an additional 63 million acres. The government also controls 528 million acres of submerged lands on the nation's Outer Continental Shelf (OCS). Onshore, 103 million acres of federally owned lands (approximately 13 percent) are now under lease for oil and gas development. Offshore, 17.9 million acres of federal lands (less than 4 percent) have been leased for oil and gas exploration. Only 2 percent of those offshore areas are now under lease.

Federal lands are estimated to hold the following energy resources:

- 37 percent of the undiscovered crude oil;
- 43 percent of the undiscovered natural gas;
- 40 percent of the demonstrated coal resources;
- 80 percent of the recoverable reserves of oil shale; and
- 95 percent of the tar sands in Utah, where most of this resource is to be found.

These federal lands -- mostly in the western states and Alaska -- hold great promise for increasing the nation's energy supplies in the future. They also are believed to contain significant amounts of 21 of the 29 major non-fuel minerals.

Government Acquisition of Land

The federal government has always been the largest single landowner in the United States. At various times it has held title to about four-fifths of the nation's total land area.

The government acquired those lands in a variety of ways. Some of the original 13 states ceded to the central government vast unsettled areas lying to the west. The United States bought the Louisiana Territory from France, Florida from Spain and Alaska from Russia. Other lands were purchased from the independent nation of Texas and from Mexico and added through a treaty with Great Britain covering lands in the Pacific Northwest.

Over the years, more than a billion acres of federal lands were transferred into state or private ownership to encourage homesteading, agriculture, construction of schools and railroads, support of higher education and for other purposes.

Of the one-third of the nation's acreage still in federal hands, some lands are reserved for national parks, wildlife refuges, military bases and other special purposes.

But hundreds of millions of acres have never been dedicated to such specific uses, nor have they been made available for exploration for oil, gas or other energy sources. Most of the nation's high-potential areas are in that category.

Changes in Land Policies

For most of the nation's history, the federal government's land-use policies have been based on three principles:

1. Natural resources on federal lands should be developed and used to meet the needs of all Americans.

2. Individuals and privately owned companies, operating competitively and relatively unhampered by government, can do the most efficient job of developing those resources.

3. Where practical, federal lands often can support more than one activity at a time. This is the principle of multiple use. Activities which have flourished on federal lands include mining, grazing of sheep and cattle, timber harvesting and recreation. Similarly, wildlife preservation and forest management have continued alongside oil and gas operations.

In the past 10 to 15 years, however, these basic policies have been changed through the passage of several laws and by changes in administrative procedures. The government has moved away from the principle that natural resources on federal lands should be developed for the benefit of the entire nation. Federal agencies have abandoned the principle of multiple use and have set aside massive tracts of land for single-purpose uses.

Again, some of these specialized land designations have been made for reasons which all sensible Americans would endorse -- to preserve unique areas such as the Grand Canyon and Yellowstone National Park, for example, or to protect other areas of unusual beauty or historical significance.

But recently there has been a trend toward closing off enormous amounts of land that have no special scenic or historical value, without first considering the potential value of those lands for the production of energy or other resources.

Many of these changes have taken place without government policy-makers being completely aware of all the facts. For years no section of the government kept a record of the cumulative impact of these actions on resource development.

Within the past four years, however, separate studies by the Interior Department and the Congressional Office of Technology Assessment addressed the question of how much federally owned land was available for mineral exploration and production. Although the two agencies used somewhat different approaches, each of them concluded that about two-thirds of the federal lands were formally closed, highly restricted or moderately restricted from mineral operations.[1]

For example, the Interior Department's Task Force on Availability of Federally Owned Lands (using 1974 data) found that 312.5 million acres were formally withdrawn and another 183 million acres were under such severe restrictions as to constitute de facto withdrawals. The two numbers add up to 495.5 million acres -- an area nearly five times the size of California.

To date, no one has published a more complete or accurate picture than the ones contained in those two government reports. But since the enactment of the Federal Land Policy and Management Act in 1976, the amount of land placed off limits to resource development has increased even more through programs of the U.S. Forest Service and the Bureau of Land Management. Acting in compliance with that law, the two agencies are considering millions of acres of federal land for suitability as "wilderness" areas. The studies are taking several years to complete, during which time new mineral exploration and development activities are, for the most part, banned from lands under consideration. Once final decisions are made, lands classified as wilderness are closed to roads, motorized vehicles, structures of all kinds -- and, obviously, to energy development.

Alaska contains 365 million acres of land, which is more than twice the total acreage of Texas. The U.S. Geological Survey estimates that onshore Alaska has more than 161 million acres of sedimentary rock -- the kind of formation in which virtually all oil and gas discoveries have been made.[2] But outside the Prudhoe Bay and Cook Inlet areas, less than 150 wells have been drilled to test Alaska's petroleum potential since 1900. Most of the potentially productive areas have been tested sparingly or not at all.

Despite the fact that little is known about the real petroleum potential of Alaska, Congress is considering legislation which would place approximately one-third of that state into categories which would permit no energy development. The executive branch of the federal government has already withdrawn more than 100 million acres of Alaska land -- an area larger than California -- by designating them as national monuments, wildlife refuges or natural resource areas. Most of the lands covered by executive order would also be covered by the proposed legislation.

Such decisions are being made without knowledge of whether significant oil and gas resources might be on those lands -- information which only careful drilling can provide.

A strong case can be made that it would be in the nation's best interests to find out what energy resources lie beneath the federal lands before decisions are reached about the future status of those lands.

Where lands have not been formally withdrawn from use, in many instances they are governed by so many restrictions that energy-related operations are impossible. Some of today's environmental laws, although they were not intended to be land-control legislation, are having the effect of limiting exploration and production on both federal and private lands.

Facts Needed for Future Planning

The companies that explore for oil and natural gas believe they can find and produce a great deal more energy. Some analysts have estimated the future output of oil and natural gas in the United States at 40 times current annual production levels. Some estimates are higher, while others are lower. One reason for these honest differences of opinion is that even the "experts" do not have all the facts. Reliable information on oil and gas resources can only be provided by exploration and drilling, and vast areas controlled by the federal government have never been opened for energy exploration.

If large new supplies can be found, all Americans will share the benefits of more secure energy resources, more stable energy prices, a sounder economy and a more solid foreign policy. If, on the other hand, discoveries fall below expectations, the people of this country will know that they must move more quickly to develop alternative energy sources.

Several years may be required to find and develop new oil and gas reserves in frontier areas such as Alaska, the Rocky Mountains and the Outer Continental Shelf. Therefore, it is essential that unnecessary delays be eliminated as promptly as feasible. It will take time to rebuild shrinking reserves and to stabilize oil and gas production.

Access and Environment Are Compatible

One of the most important steps the federal government can take to improve this country's energy supply situation is to provide improved access to those federal lands, onshore and offshore, which appear to have the greatest energy potential. Increased exploration of those areas can be compatible with environmental protection.

Petroleum operations are often temporary in nature and they disturb relatively small areas of land. Seismic surveys and test drilling, for example, involve minimal surface disturbance of the land over a fairly short period.

An oil-well site rarely covers more than a few acres. Several wells usually can provide enough information to assess the potential of a large area.

If a well proves to be a dry hole, the rig is removed and reclamation begins. If a commercial discovery is made, the normal productive life of a field is usually 25 to 30 years. Production facilities involve only small-scale disturbances that can be substantially or entirely corrected after production ceases.

Furthermore, state and federal agencies monitor every phase of petroleum operations carefully to assure that all applicable environmental regulations are obeyed.

Experience in many areas has proved that petroleum operations can be compatible with many other land uses, including wildlife management. This has been demonstrated from the Kenai Moose Range and the caribou grazing grounds in Alaska to the wildlife refuges along the coast of the Gulf of Mexico.

The Wildlife Refuge Management Act specifically states that oil and gas operations may proceed on wildlife refuges. For several years, however, federal administrators have refused to accept applications for leases in those areas, thus creating *de facto* withdrawal.

On the North Slope of Alaska, wildlife have adapted easily to the presence of men and their equipment. Caribou, moose and their calves can be seen grazing within a few hundred yards of drilling rigs, or resting and browsing beneath the elevated portions of the trans-Alaska oil pipeline, because it is warmer there and the grass is literally greener.

When carried on in an environmentally responsible manner, petroleum exploration and production need not harm animal habitats nor threaten other desirable values. Zero environmental impact is not possible, of course; a solitary camper will in some way affect the environment. But limited temporary effects from careful exploration may in some instances be a bargain price to pay for producing energy that cannot be obtained in any other way.

A Source of Revenue for the Government

It is sometimes overlooked that petroleum development can also be an important source of revenue for the federal government.

From 1953 through 1979, according to Interior Department reports, the total value of oil, gas and related products from federal OCS operations exceeded $49.7 billion. In addition to spending billions of dollars exploring and drilling offshore, the oil companies paid to the federal government more than $34 billion in cash bonuses, rentals and royalties. Thus -- even before taxes -- the government received 69 percent of the value of production from wells on federal OCS leases.

During 1979, the companies paid the federal government cash bonuses totaling $4.6 billion in five OCS lease sales. They also offered more than $1 billion in high bonus bids to the state of Alaska and the federal government for tracts in the Beaufort Sea, off Alaska's northern coast. No bids have been accepted pending court action on two lawsuits seeking to cancel the sales. Until the courts decide the issues, the high bidders will not know whether they will be allowed to explore for oil and gas in the Beaufort Sea.

Delays Slow Offshore Exploration

Since drilling began in federal waters in the 1950s, the search has been subjected to many delays. Federal leasing has been a stop-and-go affair. Many areas believed to have great potential have not been offered for leasing. The program has been plagued by cancellations and postponements of scheduled sales. Many of the delays were caused by lawsuits which eventually were thrown out of court. Such delays cost American consumers valuable time and money and increase this country's dependence on imported oil.

Here are some examples of delays:

- In August 1976, companies paid the government $1.1 billion in bonuses for leases in the Baltimore Canyon area off the mid-Atlantic coast. Because of court challenges, the successful bidders had to wait 18 months before they could start drilling. The box score in the Baltimore Canyon as of early June 1980: three potential discoveries, 18 wells which were either dry holes or found to have no significant amounts of oil and gas, and two wells still being drilled. The companies have invested more than $1.6 billion in that area, with no assured commercial discoveries.

- A lawsuit forced cancellation of a January 1978 lease sale covering portions of the Georges Bank area, off New England. The sale was rescheduled for December 1979. At that time the companies paid the government $816.5 million in cash bonuses. Indications are that by the time all necessary government permits are obtained, all plans approved by the government and all legal obstacles removed, drilling may begin in the spring or summer of 1981.

Even after leases are purchased and all permits obtained, offshore drilling often produces disappointing results. Here are some examples:

- In 1973, oil companies paid the government nearly $1.5 billion in cash bonuses for leases in the eastern Gulf of Mexico, where scientists thought there was high potential. They drilled 18 holes at a cost of several million dollars without finding producible oil and gas.

- In 1976, the companies paid the government nearly $560 million for leases in the northern Gulf of Alaska. The 11 wells drilled in that area to date have all been dry holes.

- In 1978, the government collected nearly $101 million for leases in the Southeast Georgia Embayment, off the South Atlantic coast. So far, the companies have drilled six dry holes there at a total estimated cost of $19 million.

Congress passed new legislation affecting OCS leasing and drilling late in 1978. The law required the preparation of some 40 new sets of regulations, some of which have not yet been completed. Under the new law, offshore operations are now even more complex and time-consuming than before.

The Interior Department has prepared a new five-year OCS leasing schedule which offers more hope of reducing oil imports than earlier proposals. But there is still room for further improvement in the system if we are to move ahead toward a 1990 objective of cutting oil imports by perhaps as much as one-half -- particularly in expediting the leasing of more Alaskan offshore areas, in lengthening leases to more than five years in frontier areas and in cutting the time required to write environmental impact statements and to obtain all the permits and approvals required by the federal and state governments.

Under the final OCS leasing schedule, the time lag between the government's call for offshore tract nominations and the announcement of a definite sale date has grown even longer than before. This planning process now requires up to 45 months in some frontier areas and up to 33 months even in proven areas such as the Gulf of Mexico. After leases are sold, much additional time is required to obtain all the necessary permits and approvals before drilling can start.

Industry spokemen have testified that unless Congress takes action to simplify and speed up the offshore leasing and permitting process, OCS oil and gas production will fall far short of what is needed to reduce United States dependence on imported oil.

Even with existing delays, federal offshore leases accounted for 9.1 percent of the crude oil and natural gas liquids and 23.5 percent of the natural gas produced in the United States in 1979. These volumes of oil and gas were produced from the 2 percent of the federal OCS currently under lease for petroleum exploration. Oilmen are confident that greater and more timely leasing of these offshore lands by the federal government can sharply increase the discovery of new reserves of oil and gas.

Onshore Federal Leasing

Delays have also taken their toll onshore. At the end of 1979 the Interior Department reported that 103 million acres of federal onshore lands were under lease for oil and gas operations. This represented only 13 percent of the 775 million acres owned by the government. Twenty years ago, 12 million more acres of federal land were under lease.

Although only a small percentage of federal onshore lands are under lease, the wells on those leases provided 7 percent of the liquid petroleum and 6.1 percent of the natural gas produced in this country in 1979. Oilmen believe those percentages can be increased significantly within the coming decade through improved access to federal lands.

Federal Coal Lands and Leasing

The estimated coal resources in the United States -- about 4,000 billion tons -- represent almost 32 percent of the estimated world coal resources. The minable United States coal reserve base is about 438 billion tons and the recoverable reserves are around 250 billion tons.

Nationwide, it has been estimated that federal lands contain some 175 billion tons of coal -- which amount to almost 40 percent of the total minable coal reserve base. However, production of coal from government lands represents only 8 percent of total U.S. coal production. Only a small portion of federal land has been leased to the private sector for coal development. These presently leased lands contain only 27 billion tons of coal. And only 16 billion tons of those 27 billion tons are currently recoverable.

The largest landowner, by far, of coal west of the Mississippi is the federal government. This area of the country contains most of the cleaner burning low-sulfur coal and therefore much new coal development is likely to occur there. Federal lands contain about 60 percent of the coal west of the Mississippi.

In addition to the lands it owns outright, the federal government also has control over an additional 20 percent of coal reserves because of "checkerboard" ownership patterns. Companies have sought to exchange tracts of land so that they could develop the coal reserves on adjacent tracts. But, so far, the companies have been unable to obtain government permission to make these exchanges.

In 1971 the Department of the Interior's Bureau of Land Management (BLM) halted leasing of all coal-bearing federal lands. It did so in order to reassess leasing policy. At that time, only about 10 percent of the then-leased federal lands were producing coal. Today, nine years later, the overall leasing moratorium is still in effect, as the BLM continues its review process.

The Department of the Interior has indicated that it plans to resume leasing of federal coal lands in January 1981, thus ending the 10-year moratorium.

However, even as the moratorium ends, other actions by the federal government hinder the production of coal. For example, the Surface Mining Control and Reclamation Act of 1977 prevents the mining of certain coal reserves because of potential environmental problems during and after mining. The Clean Air Act, as amended in 1977, also makes the use of coal more difficult.

Other government constraints on the increased production of coal, according to the National Coal Association, include:

- control of rights of way across publicly owned land, most of which are managed by the Interior and Agriculture departments; and

- slowness of federal and state agencies in completing environmental impact assessments required under the National Environmental Policy Act.[3]

The National Coal Association, the United Mine Workers of America and other mining groups have stated, "The government has an important role to play -- particularly by removing unnecessary constraints and allowing and encouraging increased use of coal."[4]

If the United States is to produce and use more of its abundant coal resources, the government must make available these federal lands containing huge deposits of coal.

Synfuels and Federal Lands

Government land use policies that now affect (and substantially restrict) development of oil, gas and coal on federal lands also affect the nation's emerging energy resources -- oil shale, tar sands, geothermal energy and synthetic fuels made from coal.

The trend toward land withdrawal will especially affect the future development of the huge deposits of oil shale. One area in the Rocky Mountains, known as the Green River Formation (covering 16,500 square miles within the states of Colorado, Utah and Wyoming), may contain as much as 600 billion barrels of potentially recoverable oil from shale -- 80 percent of it on federal lands. This amount of oil from shale is equivalent to a 200-year supply of oil imports at 1979 import levels.

* * * *

The United States can move toward a 1990 goal of increased energy security by developing and using our domestic energy resources. However, increased access to these resources is essential.

X. ECONOMIC BENEFITS OF REDUCING IMPORTS

Additional domestic energy production will improve our bargaining position with the OPEC oil cartel. By moving toward a national goal of cutting imports by as much as 50 percent by 1990, we will face a smaller risk of sudden, steep increases in world oil prices. This in turn will mean an economy with a more favorable outlook over the decade for continued economic growth, higher wages and salaries and less inflation.

Additional domestic energy, produced efficiently, should:

- increase our economy's total output of goods and services;
- weaken OPEC's ability to raise world oil prices; and
- make us less vulnerable to a sudden reduction in oil imports.

Increased Total Output of Goods and Services

Americans need energy to produce food, shelter, clothing, transportation, medical care and a multitude of other products and services. In short, energy is vitally important to the United States economy. If the United States can obtain energy more cheaply, then Americans can produce more of the products and services they need.

The best way to reduce OPEC's ability to raise the price of oil is to produce more domestic energy. By doing so the United States can reduce its demand for foreign oil while, at the same time, permitting consumers and producers to obtain the energy they need in the marketplace.

In contrast, cutting oil imports by either restricting or prohibiting the use of energy is both wasteful and expensive. It is wasteful because much time and effort are spent trying to obtain the limited amount of energy that is available. It is expensive because a great deal of output, jobs and income are lost when energy needs are not met. The net result of this course is a standard of living for Americans that falls short of its potential.

The United States can choose to produce substantially more energy using fewer resources than OPEC is now asking for its oil.

And OPEC is asking a great deal. Our tab for imported oil has grown from $26 billion in 1974, the year the embargo ended, to $60 billion in 1979. Our payments for foreign oil in 1980 are expected to approach $90 billion.

An oil import bill of $90 billion has been put into perspective by John C. Sawhill, deputy secretary of energy:

> That is more than the combined net assets of GM, GE, Ford and IBM; greater than the net income of the Fortune 500 in 1979; 2.5 times larger than what we are spending to equip the Army, Navy and Air Force. If we keep spending at this rate, we will have exported cash equal to the trading value of all stock on the New York Stock Exchange by 1990.[1]

These money figures measure the amount of time and resources that Americans must spend making the products and services necessary to pay foreign oil producers. As Richard Whalen stated in the March 1980 issue of Harper's magazine, "Whether at home or overseas, a dollar, after all, is a claim, an IOU against the U.S government."[2] OPEC agrees to accept our dollars in exchange for their oil because we have promised that, in return, OPEC can spend each and every dollar for the things we make. These products and services, provided by Americans to foreign producers, are the true cost of foreign oil.

On the other hand, we can provide additional domestic energy with less labor, capital and other resources than it takes to buy foreign oil -- if government policies that deter production of domestic energy are revised. More of our resources would be saved to help produce more housing, education, health care, clothing, food and other important consumer needs. Also, additional supplies of these products and services would contribute to lower measured rates of inflation.

Weakened OPEC Ability to Raise World Oil Prices.

Producing more energy at home will reduce domestic demand for oil imports and therefore cut total demand for OPEC oil. OPEC's pricing power is based on the demand for its product, so more energy produced in the United States can be effective in restraining OPEC from raising prices.

Replacing a substantial amount of oil imports with domestic energy would reduce sales for the OPEC cartel. History suggests that cartels are vulnerable to internal bickering and disagreement when reduced sales -- and therefore reduced revenues -- are allocated among their members. Each member tends to feel that it has been asked to shoulder too much of the loss. As a result, each member is tempted to defy the decision of the cartel and sell more than its allotted share.

By producing more domestic energy and displacing sales by OPEC, we can make it more difficult for that cartel to maintain its internal cohesion.

Since the 1973-1974 embargo, the economies of many OPEC members have become heavily dependent on oil revenues. Therefore, powerful political, economic and social forces within these countries are likely to push to maintain oil sales. But, for OPEC to maintain or increase world oil prices in the face of more competition, its members must cut back on their rates of production. Increased domestic energy production and conservation would encourage rebellious OPEC members not to follow the cartel's policy and could have a significant influence on world oil prices.

And, in fact, the historical record shows that after the 1973-1974 embargo, OPEC was not able to keep world oil prices, adjusted for inflation ("real" prices), from falling below the peak levels achieved during the embargo. It was not until the Iranian revolution five years later, and its effects on world oil supplies relative to demand, that OPEC was able to raise real prices above their peak embargo levels.

In 1978 the U.S. Department of Energy (DOE) calculated the probable effect on oil prices if the United States were to reduce its demand for imports by 1 million barrels per day. According to DOE, this could mean an oil price more than $1 per barrel less than would otherwise be the case.[3] Should the United States succeed in cutting imports in half by 1990, its demand for foreign petroleum would be cut by 4 million barrels a day.

Even modest increases in energy supply can have a considerable downward effect on price because of what economists call the relative inelasticity of demand for petroleum. This means that, although product prices affect demand, a price change leads to a smaller percentage change in the amount that users want to buy. The other side of this coin is that petroleum prices tend to be very sensitive to changes in supply. That is, a relatively small change in the amount of oil supplied can cause a much larger percentage change in price.

In the 1970s, world oil markets underwent two major shocks when supplies were less than anticipated by users: the 1973-

1974 embargo and then, later, the Iranian revolution. To continue operating their factories, homes, automobiles and other capital equipment at desired levels, users competed vigorously for the supplies that were available. In this environment OPEC countries were able to ask for -- and get -- a percentage increase in price that was several times the percentage reduction in expected oil supplies.

But an additional 4 million barrels of oil per day on the world market would create a different environment. This may appear to be a small amount compared to total world oil supplies. However, to find new buyers for the 4 million barrels per day, foreign oil producers would have to offer users more attractive terms, including lower prices, more secure contracts and easier credit terms than would otherwise be necessary.

If OPEC countries could find enough new buyers by offering terms that are only slightly better than they would otherwise extend, the United States would have little influence on price by cutting imports. But the amount of oil that users demand increases at a relatively slow pace in relationship to cuts in the asking price. This means foreign oil producers would have to lower their asking price substantially to find enough buyers who, together, would buy an additional 4 million barrels of oil per day.

If the United States cuts imports in half by 1990, there would be an additional 4 million barrels of oil a day for world oil markets. Thus, by cutting imports in half by 1990, the United States could exert substantial downward pressure on price and do a great deal to restrain future price hikes by OPEC.

More domestic energy unquestionably helps consumers obtain more energy on more favorable terms. And, consumers prefer more, not less, energy no matter how well they can adjust to tighter energy supplies. According to Professors James McKie, Michael Kennedy and E. Victor Niemeyer of the University of Texas at Austin:

> The U.S. economy compensates for a scarcity of energy in part by substituting other things for it in production and consumption, but consumers still would be better off with cheap energy than with expensive energy. The fact that we have a flexible and versatile economy cushions the blow of higher energy prices, but does not eliminate it.[4]

A specific example makes this point clear. More fuel-efficient automobiles can soften the blow of higher oil prices set by OPEC. But those same higher prices force the motorist either to replace his present car sooner than he had originally planned or else to pay more for each tankful. The consumer knows that he has been hurt by OPEC's action no matter what countermeasure he takes in response.

Permitting OPEC to decide just how scarce energy will be allows it, in effect, to determine the amount that American consumers will pay for energy. Stuart E. Eizenstat, assistant to the president for domestic affairs and policy, has said, "The amount which we pay to OPEC producers acts in the same way as a giant tax on the American consumer of energy."[5] But unlike taxes paid to federal, state and local governments, consumers receive no services in return for the "tax" paid to OPEC.

The United States need not -- and should not -- simply accept OPEC's decision on just how scarce energy will be. Our extensive energy resources can add substantially to supply -- if government policies permit and encourage their development.

Lessened Vulnerability to a Sudden Reduction in Oil Imports

By choosing to produce more domestic energy, conserve more and cut imports in half by 1990, we can become less vulnerable to a sudden reduction in oil imports. Several economic experts have cited the oil embargo of 1973-1974 -- and the sharply higher energy prices and problems of adjustment left in its wake -- as a significant cause of the severe 1974-1975 recession. Some economists consider it to be the most significant cause. And, at our current level of dependence on foreign oil, a sudden reduction in imports would exact a heavy price in terms of lost jobs, lost income and more inflation.

The embargo and the sudden jump in world oil prices in 1973-1974 provide important evidence of this vulnerability. According to Energy Future, the report of the energy project at the Harvard Business School, the oil crisis of 1973-1974 delivered "a powerful economic and political shock to the entire world. It interrupted or perhaps even permanently slowed postwar economic growth."[6] The Federal Energy Administration, now the Department of Energy, estimated that the embargo cost Americans $10 billion to $20 billion in lost income.[7] Robert Rasche and John Tatom, writing in the monthly Review of the Federal Reserve Bank of St. Louis, estimated that the increase in world oil prices engineered by OPEC in 1973-1974 reduced domestic potential output by approximately 5 percent between the fourth quarters of 1973 and 1974.[8] This loss is on the order of $60 billion (measured in 1972 dollars).

A disturbance similar to the 1973-1974 embargo would have more serious effects today since our dependence on foreign oil has grown substantially since the mid-1970s. In 1979 the United States imported more than three times as much oil from the Arab members of OPEC as in 1972, the year before the embargo. This level of dependence from a politically and socially unstable part of the world raises risks that concern thoughtful observers. The Harvard Business School's Energy Future states:

> Political instability in the Middle East, supply interruptions, the extension of Soviet influence -- such factors only make a very bad situation much worse. This point must be underlined . . . Dependence reinforces the twin vulnerabilities -- interruption of supplies and major price increases.[9]

In 1979, even without an embargo, the events in Iran precipitated sharp increases in world oil prices. The Organization for Economic Co-operation and Development (OECD) anticipates that, in view of the world oil price increases that took place in 1979 and early 1980, the outlook for 1980 among OECD countries (of which the United States is one) is a stagnant GNP, increased unemployment and a higher general price level.[10]

Stuart Eizenstat recently made a similar assessment for the United States economy. In a discussion of OPEC price increases, he stated:

> In 1979, increasing energy prices added an estimated 2.5 percent directly to inflation, depressed our rate of growth by 1.5 percent, and swelled unemployment by about a quarter of a million. Even assuming no further price increases, in 1980, energy prices are estimated to add 3 percent directly to our overall rate of inflation, reduce our rate of growth by 2 percent, and add up to 1.3 million to unemployment.[11]

Data Resources Incorporated (DRI) recently estimated what the United States could expect if in the 1980s imported oil prices suddenly and sharply increased. The DRI study predicts that sharp increases in prices for oil imports would result in "considerable cost in terms of lower growth of potential output, worsened inflationary performance and real income loss."[12]

More domestic energy production helps guard against these dangers in two ways: first, by making it more difficult for OPEC to suddenly and sharply raise prices; second, by increasing our energy self-sufficiency. Again, the United States is fortunate. It currently produces about 80 percent of its energy needs. Through conservation and the production of additional domestic energy, this nation could move towards providing 90 percent of its energy needs by 1990. Obviously, a sudden disruption in oil imports is less costly for a nation that is 90 percent energy self-sufficient than for a nation that is 80 percent self-sufficient. Fewer jobs are in jeopardy. Less real income and output would be lost.

In short, increased domestic energy should improve our economic stability -- and that stability is a vital part of our national security.

Conclusion

The United States need not -- and should not -- simply accept other nations' decisions about the availability and price of energy. Our extensive energy resources, if we choose to develop them, can add substantially to supply. By exercising that choice, Americans can restore a great deal of their economic security. Not only would the outlook for economic growth and less inflation be more favorable, but Americans would have more control over decisions and events that affect their well-being.

XI. ENERGY SECURITY -- INDIVIDUAL BENEFITS

Energy matters tend to be viewed in Washington as national economic and foreign problems with broad industrial and governmental solutions. However, those same problems are seen by each individual American to be intensely personal.

Individuals know that an energy shortage can affect their jobs. They have seen that foreign potentates can have more control than seems right over their home heating and personal transportation. And they are concerned that events over which they seem to have little control can force them to change the way they live in ways they deeply resent.

Over-reliance on imported oil does indeed affect jobs, personal freedom of choice and lifestyles. In fact, the individual must bear the ultimate burden of America's energy problems.

But each individual can regain some of the control he has lost over his own life if the nation does meet a goal of reducing oil imports by perhaps as much as 50 percent by 1990. Each person will benefit from the increased energy security of the nation.

People's jobs will be more secure. There will be less chance of layoffs caused by sudden reductions in energy supplies. More domestic energy production can make each person less vulnerable to a foreign oil embargo and the economic shock and setback that would follow.

More goods and services will be available. If the United States can produce domestic energy with less labor, capital and other resources than it takes to buy foreign oil, more of our resources will be saved for other uses. Thus, each person will be better able to meet housing, education, health care, clothing, food and other important needs.

Inflation can be lower. Over the past several years, sudden, sharp increases in world oil prices have helped raise the rate of inflation in the United States. But with additional domestic energy production, we can either avoid or moderate oil price increases that OPEC would impose.

Increases in the prices of fuel -- at home, at the pump -- can be restrained. The more energy we produce in the United States, the more we add to the world's energy supplies. More energy produced in the United States can be effective in restraining OPEC from raising prices.

Prices of imported consumer goods will be held down. The United States will import less oil and send fewer dollars abroad. When the dollar is stronger, the consumer pays less for many imported products.

Personal lifestyles will be more secure. Americans need not -- and should not -- simply accept OPEC's decision on just how scarce energy will be. With less imported energy, each person will face less risk that a disruption will trigger governmental controls that lead to long lines at service stations. Individuals will face less risk of forced changes in lifestyles -- in driving, in homes and at work.

Thus, just as individuals have had to bear the ultimate burden of the nation's energy problems, they will benefit if we begin to solve those problems. And the solutions are not outside our grasp. By acting wisely and choosing to develop domestic resources more efficiently, the United States can reduce its dependence on foreign oil by as much as 50 percent by 1990. This is a reasonable, attainable goal that will provide benefits for Americans as individuals and for this nation as a whole.

FOOTNOTES

SUMMARY

1. The sources in this table are listed on pages 158-160 of the Bibliography.
2. The sources in this table are listed on pages 158-160 of the Bibliography.
3. The sources for the 1990 base forecasts and accelerated case are listed on pages 158-160 of the Bibliography.

CHAPTER I. INTRODUCTION AND DOCUMENTATION

1. Jimmy Carter, "Remarks of the President at a Briefing on Energy Conservation in Transportation," Office of the White House Press Secretary, 29 April 1980, p. 1.
2. "U.S. Petroleum Industry Will Face Monumental Task in Next Decade," Oil & Gas Journal, 12 November 1979, p. 176.
3. Exxon Company, U.S.A., Energy Outlook 1980-2000, December 1979.
4. The sources in this table are listed on pages 158-160 of the Bibliography.
5. The sources in this table are listed on pages 158-160 of the Bibliography.
6. Edward W. Erickson, "The U.S. Supply of Crude Oil: A Discussion of the Intensive Margin and Graduated Production Response," Appears in the report of the Oil Policy Task Force of the Scientists Institute for Public Information (report is as yet untitled). New York: SIPI, 1980 (planned date of publication), p. 14 and p. 29.
7. National Academy of Sciences, Study of Nuclear and Alternative Energy Systems, U.S. Energy Supply Prospects to 2010, Washington, D.C., 1979, p. 11.
8. "U.S. Petroleum Industry Will Face Monumental Task in Next Decade," p. 176.
9. National Petroleum Council, information discussed at open meetings of the Coordinating Subcommittee of the Refinery Flexibility Committee (Washington, D.C., 11 April 1980 and 27 May 1980).
10. U.S., Department of Energy, Analysis Report, Energy Supply and Demand in the Midterm: 1985, 1990 and 1995, Washington, D.C., April 1979, DOE/EIA-0102/52, Order no., 476, p. 23.
11. Charles W. Duncan, Posture Statement Before the Committee on Science and Technology, U.S. House of Representatives, Washington, D.C., 31 January 1980, p. 6.

12. Exxon Company, U.S.A., Energy Outlook 1980-2000, p. 16.
13. Shell Oil Company, Demand/Supply Table, "National Energy Outlook, 1980-1990: An Interim Report (May, 1980)," materials distributed at 4 June 1980 press conference, Washington, D.C.
14. Robert H. Nanz, Vice President, Shell Oil Company; Statement Before Ad Hoc Select Committee on Outer Continental Shelf on the proposed five-year OCS leasing schedule, 1 August 1979, p. 1.
15. Ibid., p. 3.
16. National Coal Association, NCA Economics Committee, Long Term Forecast 1980-1985-1990, 22 February 1980, Table I.
17. U.S., Department of Energy, Energy Supply and Demand in the Midterm: 1985, 1990, and 1995, p. 23.
18. National Academy of Sciences, Study of Nuclear and Alternative Energy Systems, p. 14.
19. Ibid., p. 15.
20. U.S., Department of Energy, Analysis Report, Commercial Nuclear and Uranium Market Forecasts for the United States and the World Outside Communist Areas, DOE/EIA 0184/24, Order no. 556, Washington, D.C., January 1980, p. 40.
21. Atomic Industrial Forum, Inc., "Nuclear Power Plants in the U.S.," INFO News Release, Washington, D.C., 31 December 1979, p. 7.
22. National Electric Reliability Council (NERC), 1980 Summary of Projected Peak Load, Generating Capability, and Fossil Fuel Requirements for the Regional Reliability Councils of NERC, Princeton, New Jersey, July 1980, Tables 7-9.
23. National Academy of Sciences, Study of Nuclear and Alternative Energy Systems, p. 18.
24. Ibid., p. 23.
25. Ibid., pp. 20-21.
26. Exxon Company, U.S.A., Energy Outlook 1980-2000, p. 6.
27. Shell Oil Company, Demand Supply Table, "National Energy Outlook, 1980-1990."
28. U.S., Department of Energy, Energy Supply and Demand in the Midterm: 1985, 1990, and 1995, p. 23.
29. Charles W. Duncan, Posture Statement Before the Committee on Science and Technology, p. 5.
30. Robert Stobaugh and Daniel Yergin, "A New Look at the Energy Issue," The Washington Star, 17 February 1980, p. 6-I.
31. Engineering Societies Commission on Energy Inc., "Coal Technologies Market Analysis," prepared for U.S. Department of Energy under contract no. Ef-77-C-01-2468 (Washington, D.C., January 1980), p. IV.
32. Ibid., p. iii.

33. Engineering Societies Commission on Energy, Inc., "Coal Technologies Market Analysis," pp. IV-V.
34. The sources for the 1990 base forecasts and accelerated case are listed on pages 158-160 of the Bibliography.

CHAPTER II. NATIONAL SECURITY

1. Carter, "Remarks at a Briefing on Energy Conservation in Transportation," p. 1.
2. U.S., Department of the Treasury, Report of Investigation Under Section 232 of the Trade Expansion Act of 1962, 19 U.S.C. 1862, as Amended, 12 March 1979, p. 3.
3. Ibid., p. 3.
4. U.S., Department of the Treasury, Treasury Releases Report on the National Security Effects of Oil Imports, News Release, 21 March 1979, p. 2.
5. W. Michael Blumenthal, Memorandum to the President on Report of Section 232 Investigation on Oil Imports, 12 March 1979.
6. Edward W. Erickson, "The Strategic Military Importance of Oil," Current History, July-August, 1978, pp. 2-3.
7. U.S., Department of the Treasury, Report of Investigation Under Section 232 of the Trade Expansion Act of 1962, p. 7.
8. "Prices May Weaken West's Defense, Brown Warns Oil Countries," The Washington Post, 21 June 1980, p. A-10.
9. Central Intelligence Agency, Statement of Admiral Stansfield Turner before the Committee on Energy and Natural Resources, U.S. Senate, 22 April 1980, p. 13.
10. U.S., Department of the Treasury, Report of Investigation Under Section 232 of the Trade Expansion Act of 1962, p. 7.
11. George Marienthal, Deputy Assistant Secretary of Defense for Energy, Environment and Safety; Remarks Before the Subcommittee on Military Construction of the Committee on Appropriations, 24 March 1980.
12. M.A. Adelman et al., Oil, Divestiture and National Security, (New York: Crane, Russak & Company, Inc., 1977), p. 187.

CHAPTER III. OIL AND NATURAL GAS

1. American Petroleum Institute, News Release (Washington, D.C., 8 May 1980); American Gas Association, News Release (Washington, D.C., 5 May 1980).

2. Charles D. Masters, "Recent Estimates of U.S. Oil and Gas Resource Potential," Speech before the Annual Meeting of the American Association for the Advancement of Science (Houston, Texas, 5 January 1979), p. 2.
3. U.S. Geological Survey, Geological Estimates of Undiscovered Recoverable Oil and Gas Resources in the United States, Geological Survey Circular 725 (1975), pp. 26-35, p. 45.
4. U.S., Department of the Interior, "Preliminary Revised Estimates of OCS Oil and Gas Resources," News Release (Washington, D.C., 7 March 1980).
5. U.S. Geological Survey, "Oil and Gas Prospects Brighten for Mid-Atlantic OCS Lease Area," News Release (Washington, D.C., 16 May 1980).
6. U.S., Department of the Interior, "Revised Estimates of OCS Oil and Gas Resources."
7. Jimmy Carter, "Oil and Gas Development Program for the National Petroleum Reserve in Alaska," Office of the White House Press Secretary (28 January 1980), p. 1.
8. Philip K. Verleger, Jr., "Thwarting Energy Independence," Harper's, April 1980, p. 111.
9. Hans H. Landsberg et al., Energy: The Next Twenty Years, Report by a study group sponsored by the Ford Foundation and administered by Resources for the Future (Cambridge: Ballinger Publishing Company, 1979), p. 386.
10. "U.S. Industry Spending to Hit Another Record," Oil & Gas Journal, 18 February 1980, p. 55.
11. Ibid., p. 56.
12. U.S., Department of Energy, Energy Information Administration, "Cost and Indexes for Domestic Oil Field Equipment and Production Operations in the United States," DOE/EIA-0096, October 1978, p. 10.
13. "Hughes Rig Count," Oil & Gas Journal, 23 June 1980, p. 208.
14. "Price Decontrol for Natural Gas Found at 15,000 Feet or More Spurs Exploration," The Wall Street Journal, 27 March 1980, p. 46.
15. "EOR Methods Help Ultimate Recovery," Oil & Gas Journal, 31 March 1980, p. 79, p. 80.
16. Ibid., p. 116.
17. U.S., Department of Energy, "Projections of Enhanced Oil Recovery, 1985-1995," DOE/EIA-0183/11, September 1979.
18. National Science Foundation, "Research and Development in Industry," (Washington, D.C., 1977).
19. U.S., Department of Commerce, Office of Technology Assessment and Forecasting, Patent and Trademark Office, "Patent Profiles: Synthetic Fuels," (Washington, D.C., December 1979).
20. James S. Cannon and Stewart W. Herman, Energy Futures: Industry and the New Technologies, (New York: INFORM, Inc., 1976).

21. Comptroller General of the United States, "Analysis of Current Trends in U.S. Petroleum and Natural Gas Production," report to the Congress of the United States, (Washington, D.C., 7 December 1979).

CHAPTER IV. COAL

1. National Coal Association, "Coal Leasing -- Interior's New Federal Coal Management Program (FCMP)," Coal Policy Issues.no. 22 (Washington, D.C.; 28 March 1980).
2. Carl E. Bagge, Speech to the Denver Coal Club, Denver, Colorado, 14 February 1980, p. 4.
3. Ibid.
4. Ibid., p. 5.
5. Comptroller General of the United States, "Water Supply Should Not Be an Obstacle to Meeting Energy Development Goals," Report to the Congress (24 January 1980), p. i.
6. U.S., Department of the Interior, Bureau of Land Management, Final Environmental Statement Federal Coal Management Program, April 1979, pp. 5-78.
7. R.E. Samples, Statement Before the Energy Subcommittee of the Republican Platform Committee, Detroit, Michigan, 8 July 1980, p. 4.
8. Interview with Denny Ellerman, Director of Policy Analysis and Evaluation, National Coal Association, Washington, D.C., 11 July 1980.
9. U.S., Department of Energy, Economic Regulatory Administration, Powerplant and Industrial Fuel Use Act Annual Report, 1 March 1980, p. 40.
10. The President's Commission on Coal, Staff Findings, Washington, D.C., March 1980, p. 2.
11. National Coal Association, "Rail Carrier Regulation," Coal Policy Issues no. 17A (Washington, D.C.: 30 April 1980).
12. Robert Stobaugh and Daniel Yergin, eds., Energy Future, Report of the Energy Project at the Harvard Business School (New York: Random House, 1979), p. 88.
13. Carroll L. Wilson, Project Director, Massachusetts Institute of Technology, Coal -- Bridge to the Future, Report of the World Coal Study (Cambridge: Ballinger Publishing Company, 1980) p. xvii.
14. Ibid., p. 40.
15. Comptroller General of the United States, "Water Supply Should Not Be an Obstacle to Meeting Energy Development Goals," p. iii.
16. Douglas M. Costle, Administrator, Environmental Protection Agency, Statement Before the Subcommittee on Environment and Public Works, U.S. Senate (Washington, D.C., 19 March 1980) p. 4.

17. Electric Power Research Institute, "Acid Rain," Energy Researcher (Palo Alto, California; Spring, 1980), p. 1.
18. Ibid.
19. "Storm Brewing Over Acid Rain Effects," The Journal of Commerce, 9 April 1980, p. 1.
20. William N. Poundstone, "Let's Get the Facts About Acid Rain," Speech before the 1980 American Mining Congress International Coal Show (Chicago, Illinois, 9 April 1980), p. 1.
21. Ibid.
22. Ibid., p. 4.
23. William N. Poundstone, Statement on Behalf of the National Coal Association, American Mining Congress, Bituminous Coal Operators Association and the National Independent Coal Operators Association before the Senate Small Business Subcommittee (Washington, D.C., 14 March 1979), p. 6.
24. Electric Power Research Institute, "Acid Rain," p. 4.
25. Poundstone, "Facts About Acid Rain," p. 9.
26. Wilson Quarterly (Washington, D.C., Spring 1980), pp. 43-44.
27. Wilson, Coal -- Bridge to the Future, p. xvii.
28. Douglas M. Costle, Statement presented to the Subcommittee on Energy Regulation, U.S. Senate Committee on Energy and Natural Resources (Washington, D.C., 23 April 1980).
29. Barbara Blum, Statement before the Subcommittee on Energy and Power, Committee on Interstate and Foreign Commerce, U.S. House of Representatives (Washington, D.C., 2 April 1980).
30. The President's Commission on Coal, Recommendations and Summary Findings, John D. Rockefeller IV, Chairman (Washington, D.C.: Government Printing Office, 3 March 1980), p. 9.
31. Wilson, Coal -- Bridge to the Future, p. xvii.
32. United Mine Workers of America, National Coal Association, Bituminous Coal Operators Association and American Mining Congress, Letter to the President of the United States, 6 March 1980, p. 7.
33. Energy Laboratory, e-lab (Massachusetts Institute of Technology, April-June 1979), p. 1.
34. "More Coal Per Ton," EPRI Journal (June 1979), p. 7.
35. Kurt E. Yeager, Director, Fossil Fuel Power Plants Department, Electric Power Research Institute, in editorial, "Cleaning Up Coal," EPRI Journal (June 1979), p. 2.
36. "Particulates/Control: Clearing the Air," EPRI Journal (October 1978), p. 33.
37. e-lab, p. 4.
38. Foster Wheeler Energy Corporation, "NEWS" (Livingston, New Jersey, 15 November 1979), p. 1.
39. e-lab, p. 6.
40. Samples, Statement Before the Republican Platform Committee, p. 10.

41. Carl E. Bagge, Letter to the President of the United States (7 May 1980), p. 2.
42. Wilson, Coal -- Bridge to the Future, p. xvii.

CHAPTER V. NUCLEAR ENERGY

1. Jimmy Carter, "Remarks of the President on the Kemeny Commission Report on Three Mile Island," Office of the White House Press Secretary (7 December 1979).
2. Atomic Industrial Forum, Inc., "Nuclear Power Plants in the U.S.," p. 7.
3. National Electric Reliability Council, 1980 Summary of Projected Peak Load, Generating Capability, and Fossil Fuel Requirements Tables 7-9.
4. U.S., Department of Energy, Statistical Data of the Uranium Industry (Washington, D.C., January 1979).
5. Nuclear Energy Policy Study Group, Nuclear Power Issues and Choices, sponsored by the Ford Foundation (Cambridge: Ballinger Publishing Company, 1977), p. 74.
6. "A Tale of Two Wastes," Commentary, 66 (November 1978), p. 64.
7. Ibid.
8. Gary Dav and Robert Williams, "Secure Storage of Radioactive Wastes," EPRI Journal, 1 (July - August 1976), pp. 6-7.
9. Atomic Industrial Forum, Inc., "Nuclear Reactor Licensing" (Washington, D.C., May 1979).
10. The President's Commission on the Accident at Three Mile Island, The Need for Change: The Legacy of TMI, John G. Kemeny, Chairman (Washington, D.C., October 1979), p. 34.
11. New Hampshire Times, 30 January 1980.
12. Subcommittee on Energy Research and Production, Committee on Science and Technology, Nuclear Powerplant Safety After Three Mile Island, U.S. House of Representatives, 97th Congress, 2d session (March 1980, Preliminary Print), p. 8.
13. Atomic Industrial Forum, Inc., "The Nuclear Industry in 1979: Applying the Lessons of the Three Mile Island Accident, Gaining New Recognition for Its Vital Energy Contribution" (Washington, D.C., 14 January 1980), p. 2.
14. Ibid., p. 3.

CHAPTER VI. SYNTHETIC FUELS AND RENEWABLE ENERGY

1. Comptroller General of the United States, Report to the Congress of the United States: Conversion of Urban Waste to Energy: Developing and Introducing Alternate Fuels from Municipal Solid Waste, EMD 79-7, 28 February 1979, p. I-2.
2. Ibid., p. 6-2.

3. Ibid., p. ii.
4. U.S., Department of Energy, Monthly Energy Review, February 1980, p. vi.
5. U.S. Congress, House, Findings and Recommendations of the Advisory Panel on Synthetic Fuels to the Committee on Science and Technology, 31 January 1980, p. 36.
6. Ibid., p. 36.
7. U.S., Department of Energy, Health and Environmental Effects of Coal Gasification and Liquefaction Technologies: A Workshop Summary and Panel Reports for the Federal Interagency Committee on the Health and Environmental Effects of Energy Technologies, edited by Richard Brown and Alice Witter, DOE/HEW/EPA-03, MTR-79W00137, May 1979, p. 4.
8. Ibid., p. 101.

CHAPTER VII. ENERGY CONSERVATION

1. Union Carbide Corporation, American Attitudes on Conservation and Government Programs to Encourage More Efficient Use, New York, November 1979, p. 6; Laurence D. Wiseman, Vice President, Yankelovich, Skelly and White, Inc., Speech before American Petroleum Institute's General Committee of Public Relations, San Diego, 17 January 1980.
2. U.S., Department of Energy, Monthly Energy Review, 27 May 1980, p. 2.
3. Central Intelligence Agency, CIA International Energy Statistical Review, ER-IESR 80-008, 23 April 1980.
4. Stuart E. Eizenstat, Speech before the Institute of Politics, John F. Kennedy School, Harvard University, Cambridge, Massachusetts, 21 April 1980, p. 13.
5. Schurr, Energy in America's Future, pp. 137-143.
6. Ibid.
7. Ibid., pp. 133-134.
8. Ibid., p. 140.
9. Ibid., p. 127-143.
10. Ibid., p. 143-159.
11. Ibid., p. 75.
12. Ibid., p. 160.
13. Council of Economic Advisers, Economic Report of the President: Transmitted to the Congress, Washington, D.C., January 1980, p. 248.
14. U.S., Department of Energy, Annual Report to Congress 1979, Vol. II (Washington, D.C.: Government Printing Office, DOE/EIA-0173(79)/2), p. 9; U.S., Department of Energy, Monthly Energy Review, June 1980, p. 19.
15. Results of the American Petroleum Institute's Survey of Energy Conservation in Petroleum Refining for the period 1 July through 31 December 1978.

16. U.S., Department of Energy, Annual Report: Industrial Energy Efficiency Program, July 1977 through December 1978, December 1979, p. 6.
17. Eizenstat, Speech before the Institute of Politics, Harvard University, 21 April 1980, p. 13.

CHAPTER VIII. ENERGY AND THE ENVIRONMENT

1. Verleger, "Thwarting Energy Independence," p. 110.
2. Ibid., p. 111.
3. Landsberg, Energy: The Next Twenty Years, p. 30.
4. Verleger, "Thwarting Energy Independence," p. 111.
5. Eizenstat, Speech before the Institute of Politics, Harvard University, 21 April 1980.
6. "California Stiffens Rule on Smog, Sees Increase in Oil Drillers' Output," The Wall Street Journal, 10 March 1980, p. 13.
7. The President's Commission on Coal, Recommendations and Summary Findings, p. 9.

CHAPTER IX. FEDERAL LANDS -- THE NEED FOR ACCESS

1. U.S., Department of the Interior, Final Report of the Task Force on the Availability of Federally Owned Mineral Lands, Vol. I (Washington, D.C., 1977); Congressional Office of Technology Assessment, Management of Fuel and Nonfuel Minerals in Federal Lands (Washington, D.C., April 1979).
2. U.S. Geological Survey, Estimates of Oil and Gas Resources, circular 725, p. 17.
3. Carl E. Bagge, Letter to the President of the United States (Washington, D.C., 12 January 1978).
4. United Mine Workers of America, National Coal Association, Bituminous Coal Operators' Association, American Mining Congress; Letter to the President of the United States, (Washington, D.C., 6 March 1980), enclosure, p. 2.

CHAPTER X. ECONOMIC BENEFITS OF REDUCING IMPORTS

1. John C. Sawhill, Speech before the Texas Independent Producers and Royalty Owners, Houston, Texas, 29 April 1980, pp. 1-2.
2. Richard J. Whalen, "Negotiable Instruments," Harper's, March 1980, p. 25.
3. U.S., Department of Energy, An Analysis of the Impact of a One Million Barrel Per Day Demand Reduction on World Oil Prices, prepared by Mark E. Rodekohr and W. Calvin Kilgore, AM/IA/79-03, 12 October 1978, Tables 3 and 4, pp. 12-17.

4. James W. McKie, et al., "Energy and Economic Growth," Speech before the University of Texas at Austin; Austin, Texas, 29 April 1976, p. 24.
5. Eizenstat, Speech before the Institute of Politics, Harvard University, 21 April 1980, p. 4.
6. Stobaugh and Yergin, eds., *Energy Future*, pp. 4-5.
7. Federal Energy Administration, *Project Independence Report*, November 1974, p. 291.
8. Robert H. Rasche and John A. Tatom, "The Effects of the New Energy Regime on Economic Capacity, Production, and Prices," *Federal Reserve Bank of St. Louis: Review*, Vol. 59, No. 5, May 1977, p. 10.
9. Stobaugh and Yergin, eds., *Energy Future*, p. 5.
10. "Highlights from OECD Economic Outlook," *OECD Observer*, #102, January 1980, p. 29.
11. Eizenstat, Speech before the Institute of Politics, 21 April 1980, p. 4.
12. Christopher Caton and Virginia Rogers, "Energy: The Growing Risks," *The Data Resources U.S. Long-Term Review*, Spring 1979, p. 127.

BIBLIOGRAPHY

SOURCES CONSULTED FOR CHAPTER I 1990 BASE AND ACCELERATED CASES

American Gas Association. Comparison of Conventional Natural Gas Supply Forecasts. Arlington, Virginia, September 1979.

Atomic Industrial Forum, Inc. "Nuclear Power Plants in the U.S.," INFO News Release. Washington, D.C., 31 December 1979.

The Chase Manhattan Bank. The Petroleum Situation. Volume 3, August 1979.

Duncan, Charles W.; Secretary, Department of Energy. Posture Statement Before the Committee on Science and Technology, U.S. House of Representatives. Washington, D.C., 31 January 1980.

Engineering Societies Commission on Energy Inc. Coal Technologies Market Analysis. Washington, D.C., January 1980. Prepared for U.S. Department of Energy under contract No. EF-77-C-01-2468.

Erickson, Edward W. "The U.S. Supply of Crude Oil: A Discussion of the Intensive Margin and Graduated Production Response." Appears in the report of the Oil Policy Task Force of the Scientists Institute for Public Information (report is as yet untitled). New York: SIPI, 1980 (planned date of publication).

Exxon Company, U.S.A. Energy Outlook 1980-2000. December 1979.

Godley, Nigel. Arthur D. Little, Inc. Cambridge, Massachusetts. Telephone Conversation. 13 June 1980.

Maddox, Michael. Data Resources, Inc. Washington, D.C. Telephone Conversation. 13 June 1980.

Meloe, Tor. "America's Energy Future in the 1980s, What Next?" Texaco Star, Number One, 1980, pp. 22-23.

Nanz, Robert H. Statement before Ad Hoc Select Committee on Outer Continental Shelf on the proposed five-year OCS Leasing Schedule. 1 August 1979.

National Academy of Sciences. Study of Nuclear and Alternative Energy Systems, U.S. Energy Supply Prospects to 2010. Washington, D.C., 1979.

National Coal Association. NCA Economics Committee, Long Term Forecast 1980-1985-1990. 22 February 1980.

National Electric Reliability Council (NERC). 1980 Summary of Projected Peak Load, Generating Capability, and Fossil Fuel Requirements for the Regional Reliability Councils of NERC. Princeton, N.J., July 1980.

National Petroleum Council. Information discussed at open meetings of the Coordinating Subcommittee of the Refinery Flexibility Committee. Washington, D.C., 11 April 1980 and 27 May 1980.

National Petroleum Council. Information discussed at open meeting of the Refinery Capability Task Group of the Refinery Flexibility Committee. Washington, D.C., 12 May 1980.

National Petroleum Council. Refinery Flexibility, An Interim Report of the National Petroleum Council. Volume I, December 1979.

Petroleum Industry Research Foundation, Inc. Oil in the U.S. Energy Perspective -- A Forecast to 1990. New York, May 1980.

Schurr, Sam H. et al. Energy in America's Future: The Choices Before Us, A study by the staff of the Resources for the Future National Energy Strategies Project. Baltimore: The Johns Hopkins University Press, 1979.

Shell Oil Company. "National Energy Outlook, 1980-1990: An Interim Report (May, 1980)." Materials distributed at 4 June 1980 press conference. Washington, D.C.

Standard Oil Company of California, Economics Department. Data Sheet on U.S. Energy Forecast, March 1980, 801, revised 29 May 1980.

Stobaugh, Robert and Yergin, Daniel. "A New Look at the Energy Issue." The Washington Star, 17 February 1980, p. G-I.

Stobaugh, Robert and Yergin, Daniel, eds. Energy Future: Report of the Energy Project at the Harvard Business School. New York: Random House, 1979.

U.S. Congressional Budget Office. Background Paper, The World Oil Market in the 1980s: Implications for the U.S. Washington, D.C., May 1980.

U.S. Department of Energy. Commercial Nuclear and Uranium Market Forecasts for the United States and the World Outside Communist Areas, Analysis Report. DOE/EIA-0184/24, Order no. 556. Washington, D.C., January 1980.

U.S. Department of Energy. Energy Supply and Demand in the Midterm: 1985, 1990, and 1995, Analysis Report. DOE/EIA-0102/52, Order no. 476. Washington, D.C., April 1979.

U.S. General Accounting Office. Analysis of Current Trends in U.S. Petroleum and Natural Gas Production. EMD-80-24. Washington, D.C., 7 December 1979.

"U.S. Petroleum Industry Will Face Monumental Task in Next Decade." Oil & Gas Journal. 12 November 1979, pp. 170-184.

SOURCES CONSULTED FOR CHAPTERS II-XI

Adelman, M.A. et al. Oil, Divestiture and National Security. New York: Crane, Russak & Company, Inc., 1977.

American Petroleum Institute. News Release. Washington, D.C., 8 May 1980.

Atomic Industrial Forum, Inc. "Nuclear Power Plants in the U.S." INFO News Release. 31 December 1979.

Atomic Industrial Forum, Inc. "Nuclear Reactor Licensing." Washington, D.C., May 1979.

Atomic Industrial Forum, Inc. "The Nuclear Industry in 1979: Applying the Lessons of the Three Mile Island Accident, Gaining New Recognition for Its Vital Energy Contribution." Washington, D.C., 14 January 1980.

Bagge, Carl E. Letter to the President of the United States. Washington, D.C., 12 January 1978.

Bagge, Carl E. Speech to the Denver Coal Club. Denver, Colorado, 14 February 1980.

Blumenthal, Michael W. Memorandum to the President on Report of Section 232 Investigation on Oil Imports, 12 March 1979.

"California Stiffens Rule on Smog, Sees Increase in Oil Drillers' Output." The Wall Street Journal, 10 March 1980, p. 13.

Cannon, James S. and Herman, Stewart W. Energy Futures: Industry and the New Technologies. New York: INFORM, Inc., 1976.

Carter, Jimmy. "Oil and Gas Development Program for the National Petroleum Reserve in Alaska." Office of the White House Press Secretary, 28 January 1980.

Carter, Jimmy. "Remarks of the President at a Briefing on Energy Conservation in Transportation." Office of the White House Press Secretary, 29 April 1980.

Carter, Jimmy. "Remarks of the President on the Kemeny Commission Report on Three Mile Island." Office of the White House Press Secretary, 7 December 1979.

Caton, Christopher and Rogers, Virginia. "Energy: The Growing Risks." The Data Resources U.S. Long-Term Review, Spring 1979.

Central Intelligence Agency. CIA International Energy Statistical Review, 23 April 1980.

Central Intelligence Agency. Statement of Admiral Stansfield Turner before the Committee on Energy and Natural Resources. U.S. Senate, 22 April 1980.

Comptroller General of the United States. "Analysis of Current Trends in U.S. Petroleum and Natural Gas Production." Washington, D.C., 7 December 1979.

Congressional Office of Technology Assessment. Management of Fuel and Nonfuel Minerals in Federal Lands. Washington, D.C., April 1979.

Costle, Douglas M., Administrator, Environmental Protection Agency. Statement before the Subcommittee on Environment and Public Works. U.S. Senate, Washington, D.C., 19 March 1980.

Council of Economic Advisers. Economic Report of the President: Transmitted to the Congress. Washington, D.C., January 1980.

Darmstadter, Joel et al. How Industrial Societies Use Energy. Baltimore: The Johns Hopkins University Press, 1977.

Dav, Gary and Williams, Robert. "Secure Storage of Radioactive Wastes." EPRI Journal 1 (July-August 1976) pp. 6-7.

Eizenstat, Stuart E. Speech before the Institute of Politics, John F. Kennedy School, Harvard University. Cambridge, Massachusetts, 21 April 1980.

Electric Power Research Institute. "Acid Rain." Energy Researcher. Palo Alto, California; Spring, 1980.

Ellerman, Denny, Director of Policy Analysis and Evaluation. National Coal Association, Washington, D.C. Interview, 11 July 1980.

"EOR Methods Help Ultimate Recovery." Oil & Gas Journal, 31 March 1980, pp. 79-124.

Erickson, Edward W. "The Strategic Military Importance of Oil." Current History, July-August, 1978.

Federal Energy Administration. Project Independence Report, November 1974.

"Highlights from OECD Economic Outlook." OECD Observer, January 1980.

"Hughes Rig Count." Oil & Gas Journal, 23 June 1980.

Landsberg, Hans H. et al. Energy: The Next Twenty Years, Report by a study group sponsored by the Ford Foundation and administered by Resources for the Future. Cambridge: Ballinger Publishing Company, 1979.

Masters, Charles D. "Recent Estimates of U.S. Oil and Gas Resource Potential." Speech before the Annual Meeting of the American Association for the Advancement of Science. Houston, Texas, 5 January 1979.

McKie, James W. et al. "Energy and Economic Growth." Speech before the University of Texas at Austin. Austin, Texas, 29 April 1976.

National Coal Association. "Coal Leasing -- Interior's New Federal Coal Management Program (FCMP)." Coal Policy Issues, no. 2A, Washington, D.C., 28 March 1980.

National Coal Association. "Rail Carrier Regulation." Coal Policy Issues, no. 17A. Washington, D.C., 30 April 1980.

National Electric Reliability Council (NERC). 1980 Summary of Projected Peak Load, Generating Capability, and Fossil Fuel Requirements for the Regional Reliability Councils of NERC. Princeton, N.J., July 1980.

National Science Foundation. "Research and Development in Industry." Washington, D.C., 1977.

New Hampshire Times. 30 January 1980.

Nuclear Energy Policy Study Group. Nuclear Power Issues and Choices, sponsored by the Ford Foundation. Cambridge: Ballinger Publishing Company, 1977.

The President's Commission on Coal. Recommendations and Summary Findings. John D. Rockefeller, IV, Chairman. Washington, D.C., 3 March 1980.

The President's Commission on Coal. Staff Findings. Washington, D.C., March 1980.

The President's Commission on the Accident at Three Mile Island. The Need for Change: The Legacy of TMI. John G. Kemeny, Chairman. Washington, D.C., October 1979.

"Price Decontrol for Natural Gas Found at 15,000 Feet or More Spurs Exploration." The Wall Street Journal, 27 March 1980.

"Prices May Weaken West's Defense, Brown Warns Oil Countries." The Washington Post, 21 June 1980, p. A-10.

Rasche, Robert H. and Tatom, John A. "The Effects of the New Energy Regime on Economic Capacity, Production, and Prices." Federal Reserve Bank of St. Louis: Review. May 1977.

Results of the American Petroleum Institute's Survey of Energy Conservation in Petroleum Refining for the period July 1 through December 31, 1978.

Samples, R.E. Statement before the Energy Subcommittee of the Republican Platform Committee. Detroit, Michigan, 8 July 1980.

Sawhill, John C. Speech before the Texas Independent Producers and Royalty Owners, Houston, Texas, 29 April 1980.

Schurr, Sam H. et al. Energy in America's Future: The Choices Before Us, a study by the staff of the Resources for the Future National Energy Strategies Project. Baltimore: The Johns Hopkins University Press, 1979.

Stobaugh, Robert and Yergin, Daniel, eds. Energy Future, Report of the Energy Project at the Harvard Business School. New York: Random House, 1979.

"Storm Brewing Over Acid Rain Effects." The Journal of Commerce, 9 April 1980.

Subcommittee on Energy Research and Production, Committee on Science and Technology. *Nuclear Powerplant Safety After Three Mile Island*. U.S. House of Representatives, 97th Congress, 2d sess. March 1980, Preliminary Print.

"A Tale of Two Wastes." *Commentary*, 66 (November 1978), pp. 63-65.

Union Carbide Corporation. *American Attitudes on Conservation and Government Programs to Encourage More Efficient Use*. New York, November 1979.

United Mine Workers of America, National Coal Association, Bituminous Coal Operators' Association, American Mining Congress. Letter to the President of the United States, enclosure. Washington, D.C., 6 March 1980.

U.S. Congress. House. *Findings and Recommendations of the Advisory Panel on Synthetic Fuels to the Committee on Science and Technology*, 31 January 1980.

U.S. Department of Commerce, Office of Technology Assessment and Forecasting, Patent and Trademark Office. "Patent Profiles: Synthetic Fuels." Washington, D.C., December 1979.

U.S. Department of Energy. *An Analysis of the Impact of a One Million Barrel Per Day Demand Reduction on World Oil Prices*, prepared by Mark E. Rodekohr and W. Calvin Kilgore, AM/IA/79-03, 1978.

U.S. Department of Energy. *Annual Report: Industrial Energy Efficiency Program*, July 1977 through December 1978. Washington, D.C., December 1979.

U.S. Department of Energy. *Annual Report to Congress 1979*, Vol. II. Washington, D.C.

U.S. Department of Energy. Economic Regulatory Administration. *Powerplant and Industrial Fuel Use Act Annual Report*, 1 March 1980.

U.S. Department of Energy. Energy Information Administration. "Cost and Indexes for Domestic Oil Field Equipment and Production Operations in the United States." Washington, D.C., 1978.

U.S. Department of Energy. *Health and Environmental Effects of Coal Gasification and Liquefaction Technologies: A Workshop Summary and Panel Reports for the Federal Interagency Committee on the Health and Environmental Effects of Energy Technologies*, edited by Richard Brown and Alice Witter, Washington, D.C., 1979.

U.S. Department of Energy. *Monthly Energy Review*, February 1980.

U.S. Department of Energy. *Monthly Energy Review*, May 1980.

U.S. Department of Energy. *Monthly Energy Review*, June 1980.

U.S. Department of Energy. "Projections of Enhanced Oil Recovery, 1985-1995." Washington, D.C., September 1979.

U.S. Department of Energy. *Statistical Data of the Uranium Industry*. Washington, D.C., January 1979.

U.S. Department of the Interior, Bureau of Land Management. *Final Environmental Statement Federal Coal Management Program*, April 1979.

U.S. Department of the Interior. *Final Report of the Task Force on the Availability of Federally Owned Mineral Lands*, Vol. I. Washington, D.C. 1977.

U.S. Department of the Interior. "Preliminary Revised Estimates of OCS Oil and Gas Resources." News Release. Washington, D.C., 7 March 1980.

U.S. Department of the Treasury. *Report of Investigation Under Section 232 of the Trade Expansion Act of 1962, 19 U.S.C. 1862, As Amended*, 1979.

U.S. Department of the Treasury. *Treasury Releases Report on the National Security Effects of Oil Imports*. News Release, 21 March 1979.

U.S. General Accounting Office. By the Comptroller General. *Report to the Congress of the United States: Conversion of Urban Waste to Energy: Developing and Introducing Alternate Fuels from Municipal Solid Waste*, EMD 79-7, 28 February 1979.

U.S. Geological Survey. *Geological Estimates of Undiscovered Recoverable Oil and Gas Resources in the United States.* Geological Survey Circular 725, 1975.

U.S. Geological Survey. "Oil and Gas Prospects Brighten for Mid-Atlantic OCS Lease Area." News Release. Washington, D.C., 16 May 1980.

"U.S. Industry Spending to Hit Another Record." *Oil & Gas Journal*, 18 February 1980, pp. 55-59.

Verleger, Philip K. "Thwarting Energy Independence." _Harper's_, April 1980.

Whalen, Richard J. "Negotiable Instruments." _Harper's_, March 1980, pp. 24-27.

Wilson, Carroll L., Project Director, Massachusetts Institute of Technology. _Coal -- Bridge to the Future_, Report of the World Coal Study. Cambridge: Ballinger Publishing Company, 1980.

Wiseman, Laurence D., Vice President, Yankelovich, Skelly and White, Inc. Speech before American Petroleum Institute's General Committee of Public Relations. San Diego, California 17 January 1980.

4300— 8/80— 7M
4300— 8/80—10M
4300—10/80—10M

333.79 A
105331
American Petroleum Institute
.
Two energy futures :